Building a Solidarity Society

"Marianne Hill draws on her experience as an activist in both the public and private sectors of our seriously unequal society to expose and challenge the power of the wealthy, and the economic structure on which it is based. This book, then, provides its readers with invaluable insights into how to participate in the struggles for a new and just society. It is recommended, in particular, for inclusion in economics courses that all too often preclude such discussions."

—Paddy Quick, *Professor Emerita, St. Francis College and member, Union for Radical Political Economics*

"Dr. Hill as an economist for the state of Mississippi had many years of observing and commenting on the plight of its residents. This state, like much of the United States and the world, has been dominated by exploitative white capitalist supremacy throughout recent centuries. She demonstrates that communities can come together in coalitions to struggle for change, whether joining together as unions in the workplace, as rights groups or as community groups, all linked across political borders in solidarity. This book will offer an informative foundation for those struggles."

—Bill Chandler, *longtime union organizer and Executive Director of MIRA, the Mississippi Immigrants' Rights Alliance*

"Enlightening examination of the costs of the domination of U.S. politics and economics by the wealthy, interpenetrated by insightful discussion of ways forward to the qualitative, systemic change needed if we are to build an economy that truly serves people and planet. Hill puts class and power at the forefront of her analysis, and draws upon a wide variety of sources and examples to make her case that radical democracy is the way forward."

—Julie Matthaei, *Professor Emerita of Economics, Wellesley College and Boardmember, U.S. Solidarity Economy Network*

"Dr. Hill uses examples from the Hurricane Katrina disaster, the financial crisis of 2008 and the COVID pandemic to show how power can shape economic outcomes. The power relations which direct policy are typically overlooked in more mainstream economic texts. Hill critiques the theories found in those texts from her heterodox perspective. With an eye to the future, she outlines a feasible path to a more just and egalitarian economic system, and presents key features of that system. Those working for such a society will find useful references from current leaders of the movement for a solidaristic future."

—Mark Klinedinst, *Professor Emeritus of Economics, University of Southern Mississippi*

"Marianne Hill directly challenges the power structure---both economic and political---that is leading us towards the destruction of our planet. She shows what a transformed culture and economy might look like, and gives us solid guidance on how we can get there. The reader will be left with a strong sense of what needs to be done to build a caring and sustainable society."

—Susan B. Inches, *Author,* Advocating for the Environment *and former Deputy Director of the Maine State Planning Office*

"Delightfully written, this book is a gem that explains contemporary political economy to an everyday reader. Marianne Hill shows how the dynamics of recent events have been propelled by the interests of a wealthy elite. The intent and actions of this elite are critically examined using both theory and evidence. Dr. Hill's scholarly pursuits and intellectual battles with mainstream economics inform this work, making it a unique resource. She advocates for a democratically controlled market economy as a step towards a society truly dedicated to the flourishing of people and planet."

—Farida Khan, *Professor of Economics, University of Colorado*

"Marianne Hill offers a cogent and powerful discussion of the underlying fault lines of our society and provides a roadmap to move forward. The discussion starts where it must: with power, the force that pits people against each other and is blocking people's ability to realize their full potential as individuals and groups. It then points the way to move forward to construct alternatives, based on the author's own experience and her ample interaction with people actively engaged in promoting constructive alternatives contributing to build a better world for all of us."

—David Barkin, *Profesor Distinguido, Universidad Autonoma Metropolitana-Unidad Xochimilco, Ciudad de México*

Marianne T. Hill

Building a Solidarity Society

Power, People, and Planet

palgrave
macmillan

Marianne T. Hill
South Portland, ME, USA

ISBN 978-3-031-07348-9 ISBN 978-3-031-07349-6 (eBook)
https://doi.org/10.1007/978-3-031-07349-6

This Palgrave Macmillan imprint is published by the registered company Springer Nature Switzerland AG
The registered company address is: Gewerbestrasse 11, 6330 Cham, Switzerland

Preface

In the fall of 1974, I arrived at Bangladesh in the midst of a famine that claimed over a million lives. I returned to the U.S. a year later to a very different, affluent reality. I was struck hard by the fact that people around the world are living very different realities. Within communities and families, this is also true. Our understanding of these very diverse realities, so basic to how we live together, is lacking—tragically so at times. We must do better. This thought, over time, led me to write this book.

Perhaps you agree with me that our social priorities are misaligned and the consequences often dire. Transformative change, I am sure, is both urgent and feasible. I am convinced that we fail to move toward the better society that is possible, not so much because of disagreements about the facts, but because of the different theories we use to interpret them. In this book, then, I examine the mistaken theories that continue to guide us. I focus in particular on power: what it is and how we can restructure it to better serve us.

When I was teaching college classes, I found it frustrating that economics textbooks did not offer any analysis of how social and political power impacts the economy. Study of power is still considered largely off-limits in economics courses, but for those instructors who want to address the elephant in the room that is power—you have found the right book! In this text, I cover basic findings of recent cross-disciplinary research on our social and political structures. I am well aware that academic works can be inaccessible to the general reader; even simple ideas can get lost.

Ideological hegemony—what is that? I have made every effort to keep the reader engaged—I hope that social science students and the general public will find this book readable, and enjoyable.

Power in society, however, is a complex subject, and understanding how it works requires some thoughtful reading and reflection. Understanding power, of course, is only the first step toward changing it. Changing how our collective power is exercised involves reshaping traditional hierarchies and identities. This brings passions and emotions into play—it is for this reason that insights on human behavior from all the social sciences are needed when the goal is building a solidarity society.

In the first part of the book, I present facts to show that the U.S. is dominated by plutocrats and their allies who show little concern for the economic distress and insecurity suffered by so many. Then, after documenting the power of our ruling elite and the motivations that drive them, I analyze the widely shared understandings about capitalism and democracy that protect the elite and justify the power they hold. These shared understandings tell us that transformative social change is neither desirable nor feasible. History, however, tells another story; it shows us today's realities in the light of the possibilities of tomorrow.

The radical transformation of Scandinavian countries is an inspiring example from history of the kind of change that is possible. At the start of the 1900s, these countries were marked by widespread poverty and dominated by a wealthy few. Today, they enjoy some of the highest living standards in the world, made possible by governments that prioritized investing in the people. Grassroots political mobilizations—ordinary people working together under dedicated leaders—were the force behind this change.

We, too, must work together if we are to move forward. Advocates for an eco-friendly and caring society—a solidarity society—are many and, once united, can be the force that moves us irreversibly toward that imagined society. Today, however, the individualism that keeps us from relying on one another is a tradition that carries on even among advocates for solidarity; this is a roadblock to the creation of the collective power that is so sorely needed. Now, a time of unrest, is precisely the right time for building the broad coalition that is capable of successfully confronting the powerful and transforming how we live together.

This is a formidable challenge. There will be ideas better, and worse, than the ones presented in this book for accomplishing this task. As a life-long activist, I know something about how difficult this work can be.

Happily, I also know that it can be impactful and rewarding as well. I have been part of the struggles for civil rights, women's rights, the environment, and more. I have been fired because of my alternative perspectives. I am not alone in recognizing that power at the workplace, as in much of society, is seldom wielded in a way that is fair and equitable.

Over the years, I have been inspired by those who have fought the good fight with extraordinary grace and courage. I have met them wherever I have lived, be it Mississippi or Massachusetts, Bangladesh or Britain, Puerto Rico or Ohio. Their strength and courage are gifts to their communities, and they have made all the difference in my life.

As a radical feminist economist, I want to acknowledge the huge intellectual debt I owe to the Union for Radical Political Economics and the International Association for Feminist Economics. The ground-breaking work of these scholars has impacted the profession for the better. My classmates in graduate school at Yale in the Marxist study group of the 1970s also shaped my thinking, as did the Marxist-feminist group. Many in these groups have gone on to do distinguished research that makes it clear that we as a society can and must do better.

I owe a huge debt of gratitude to friends and family for their love and support over the years. A few people have read chapters as I wrote them and offered insightful feedback that sparked stimulating discussions. These include my excellent writing coach Dave Harris, my wonderful daughter Rosie Hill, Sue Inches, Carl Cappello, Joanne Croft, Mary Beth Davidson, Sandy Jorgenson, James Maier, Karin Sears, and Roberta Wentworth. My dear friend Diane Smith, niece Heather Crenshaw and guide and guru Gale Constable deserve grateful mention for their encouragement. The Bodacious Broads, my Forest Garden buddies, my Dow Woods neighbors, and my chorus were greatly helpful in getting me through these pandemic years.

Finally, I want to add that the reader will find some fascinating stories and information in the footnotes, as well as links for further reading.

South Portland, USA Marianne T. Hill

Contents

List of Tables

CHAPTER 1

Introduction: On Power, People, and Planet

What's good for business is good for the country? Not so, but what's good for people and planet is, indeed, good for the country, and for business.

In the aftermath of the COVID-19 pandemic, America's beliefs about what constitutes a good society have changed. There is a growing demand for a more egalitarian democracy and a more sustainable and just economy. Advocates from across the social spectrum insist we must transform how we do things. But too often these advocates fail to consider the formidable opposition that can be expected from those in positions of power and the obstacles presented by the deeply ingrained misconceptions that protect them. This book grapples with the issue of power and the problem of how to redistribute it so that we can move toward a society in which people and planet thrive.

Power, we know, is an inescapable reality for all of us, whether we are among the powerful or the powerless. It exists in the smallest and the largest social units, in families and in nations. It permeates decisions made throughout society.

Yet power, and the huge disparities in its distribution, are seldom a focus of the debates that set the course of our nation. But who has, or doesn't have, power is not a fact of life to be accepted without challenge. Rather, there is every reason to question today's distribution of power. In fact, we must change how we structure power if we are to reset our priorities and radically improve our collective choices. If we fail to take

M. T. Hill, *Building a Solidarity Society*,
https://doi.org/10.1007/978-3-031-07349-6_1

up this challenge, on the other hand, we risk failing our planet and future generations.

Those in power today act to protect and strengthen the system that safeguards their positions and privileges and, in doing so, they cripple our collective ability to respond adequately to even the most urgent of crises. But the future of our world is at stake. In richer nations, economic insecurity is widespread, affecting all but the wealthiest. The goal of a national prosperity that is shared by all in a spirit of community seems a distant one. In poorer countries, the challenges are heart-wrenching—simply ensuring access to clean drinking water and other basic necessities is a goal that seems unattainable. And the urgent global imperative to protect the air, water, plant, and animal species of our planet receives only inadequate acknowledgment.

What steps can enable us to reset our priorities and concentrate on the critical issues facing the world? A new configuration of power and a restructuring of our institutions is demanded; such transformative change can begin with a change in thinking, led by innovative activists.

Society is continually evolving, and activists have always been a major force shaping the future. The environmental movement, the women's movement, the Black Lives movement, the labor movement, and other social and political movements are making a difference. Activists are also behind progressive initiatives undertaken by groups like the U.N.: the U.N.'s adoption of the Universal Declaration of Human Rights was an achievement that continues to affect political debates around the globe, as does the U.N.'s Paris Agreement on climate change. In fact, people often come together to rectify perceived wrongs. Successful groups are mobilized by common sentiments and beliefs. These coalitions build a shared vision despite a multiplicity of interests. Progress, at times fleeting, is made by such groups despite the pitfalls and obstacles faced. Each step forward, however, may be only part of a much longer journey.

This book focuses on power in society and especially in economic institutions in the belief that the greater our understanding of power, the more capable we will be of taking collective action to achieve the changes needed. The work of thinkers over many years has led to a deeper understanding of power, as we point out in Chapter 2. But attitudes toward those in power vary by standpoint. Those wielding power have perspectives that differ from the views of those subject to that power. This means, for example, that employers generally view a minimum wage and other worker protections differently than do employees.

As we go about our daily lives, we are continually recreating and reshaping society. This daily collective exercise of power shapes our destiny and provides opportunities for change. Note that, in this process, the goals and desires of those with more power have more effect. Who makes decisions and how, then, requires examination—especially in light of the consequences.

Experience teaches us that new ways of doing things begin with new ways of thinking, and with the conviction that change is needed. Technological and historical developments bring new ideas—and open new opportunities. Many exciting policy innovations are being proposed in manifestos, political platforms, and other collective works today. These are influencing the on-going process of social evolution.

How do we foster the creation and spread of a commitment to positive, transformative change? This book begins with a critical look at recent crises and then examines theories that offer analyses and insights that can guide efforts for deep, positive change.

The argument laid out in the following pages builds on two basic premises. One, institutions are biased toward the preservation of existing practices and priorities. Two, perspectives that expose the most devastating consequences of the status quo clarify needed changes and help build the shared commitment that can bring them about. When different voices and viewpoints coalesce in a collective commitment to advancing a new, unifying vision, change is feasible. The American Revolution, the civil rights movement, and other collective efforts have changed the course of history. Efforts to confront pervasive injustices necessarily entail confronting those who cling to on-going ways of doing things. Those deriving power from the status quo will fiercely oppose change that means they will lose power and control, however unfulfilling their grasp on power may actually be.

Today greater weight in decision-making is needed by the many whose voices are drowned out. This is true in every socioeconomic sphere, from the family and neighborhoods through government agencies and legislative bodies to the market and financial systems. Further democratizing these institutions has the potential to bring tremendous improvements to our lives and to the outlook for future generations.

Greater democracy increases the economic and social well-being of those who gain voice. The transition is difficult; at times, violence has occurred as those excluded from full citizenship struggle to become full participants in society. However, violence is minimized in democracies that provide platforms for critical alternative standpoints to be heard. There are pitfalls and obstacles to be considered as changes are made, as well as the need to safeguard against undesired results.

We are far from agreement that a different structuring of economic and political decision-making—a more inclusive and responsible distribution of power—will lead to enhanced well-being for all of us. In fact, many of us will oppose such change.

In the end, only broad political movements are capable of shifting our course toward a more radically democratic country. There will be conflicts that need to be worked through before our leading institutions become more egalitarian, responsive, and responsible. Some of us, at least, must change our minds.

Now is the time to lay the foundation of a new egalitarian and caring society, one that embraces the values of democracy, diversity, and community. Proponents of such a caring solidarity society are many, and, once united, can be the change agents that democratize power in firms, families, and societies. Although many organizations are working toward the goal of an egalitarian and caring global society, the World Social Forum is especially notable, given the many thousands who belong to this participatory solidarity network.[1] The Forum offers its members a space for the exchange of ideas and practices, but does not make decisions or speak on behalf of its members. However, its Charter of Principles states many of the shared values of solidarity networks. These include: respect for human rights; participatory democracy; peaceful, equal, and solidarity-based relations; social justice; national sovereignty; meeting people's needs; respecting nature, and opposing the domination of the world by capital and any form of imperialism (which includes opposing globalization led by corporations and countries at their service).

This book aims to foster the clarity, cohesion, and courage that can ensure the success of such solidarity activists. We begin in Chapter 2 with an overview of three crises: the COVID-19 pandemic, the 2007–2009 financial crisis, and Hurricane Katrina. The priorities of the powerful explain their failure to respond responsibly and humanely to these crises. Power is defined, drawing on the extensive research on the subject. The power of the dominant elite is argued to be based on the institutions, identities, and shared understandings that shape social behavior. Thinkers over the years have probed the strengths and weaknesses of efforts to end the exploitation of paid and unpaid workers, and to establish a state that serves its citizens. Constraints and possibilities lie in the ideologies that shape shared understandings and in the forces that change those understandings over time. Chapter 3 considers the domination of society by the elite few. It summarizes the findings of studies that ask what the wealthy want and that examine the ways in which the wealthy are able to advance their interests to the detriment of mainstream America and the environment.

The data and insights from Chapter 3 are put into a theoretical framework in Chapter 4 that explains the hegemonic hold of the elite. The weaknesses of our current political system are linked to broader institutional weaknesses that are justified by a mix of ideological understandings. The history of the emergence of the welfare state in Europe and the U.S. makes it clear that progress is possible, and clarifies the kind of political battles that lead to positive social change. The notion of radical democracy is introduced. Once progressive political movements have gained sufficient cohesion to influence political party platforms and elections, there is the possibility of a shift in power dynamics—one that brings greater clout to the urgent demands of the mainstream. This deeper democracy has been termed "radical democracy" to make it clear that it is more representative and its leadership more accountable than is currently the case. The willingness of citizens to accept such a shift in power, though, depends on their trust in the capabilities and accountability of those challenging the status quo. This trust must be gained; it will rest upon shared experiences and shared ideologies.

Chapter 5 critiques the economic theories used to justify clinging to a capitalist market system and the domination of society by a plutocracy. The severe limitations of introductory micro- and macro-economic theory, and their importance to the ideological belief in market efficiency, are explained. Critiques of mainstream theories are summarized, focusing on their neglect of human relationships—power and caring in particular. The possibility of a new economic order that is more aligned with basic human values is made clear. Chapter 6 then asks what kind of society would best foster the thriving of people and planet. The capability approach to well-being is used to lay out what is required of such a society. A solidarity society would move beyond traditional human rights to address the quality of social relationships embedded in political, economic, and other institutions. Drawing on current initiatives around the globe, some promising possibilities are presented. The motivations that can drive political efforts to create a caring, solidarity society are laid out. Concluding remarks further address the challenges involved in uniting the multiplicity of movements and activists in coalitions capable of transformative change.

Note

1. https://ussen.org. There is a global network of Solidarity activists working collaboratively to develop ideas and practices that can move us towards solidarity societies. There are many others, not in this network, who share the common goal of sustainable, equitable societies.

CHAPTER 2

Power Matters

"Brownie, you're doing a heck of a job!" President George W. Bush to Michael Brown, head of FEMA, during the recovery from Hurricane Katrina, September 2, 2005.

The U.S. has faced several major crises since 2000. We discuss three that expose severe structural flaws in our nation's basic institutions. The botched response to Hurricane Katrina in New Orleans in 2005, the near-collapse of the financial system in 2008, and the COVID-19 pandemic all point to the need for rethinking and restructuring our political and economic systems. To do this, we must understand what went wrong, and why power was so misused. This chapter highlights events occurring during these crises that reveal the conflicting priorities at play among those making the critical decisions. We then turn to an analysis of power and the role it played in the decisions made. Staying in power is a high priority of the powerful; it is one that requires satisfying "those who count"—often at the expense of doing right. This phenomenon, in fact, goes far in explaining the behavior of decision-makers during critical moments.

M. T. Hill, *Building a Solidarity Society*,
https://doi.org/10.1007/978-3-031-07349-6_2

The COVID-19 Pandemic

The virus labeled SARS CoV-2 emerged in late 2019 in the most populous city in central China, Wuhan. It spread quickly, unleashing a global pandemic that sickened or killed millions and precipitated an economic downturn deeper than that of the Great Recession of 2007–2009. Lockdowns were imposed in many countries with little provision for ensuring that people's basic needs would be met and, as a result, social unrest spread along with the disease.

Massive fires, floods, and/or droughts also afflicted every continent except Antarctica during the pandemic years. Adding to the turmoil of the time, videos of police killings of unarmed Black men and women in the U.S. fueled demonstrations there against racist police brutality and sparked Black Lives Matter protests around the globe. Still, above all else, the pandemic, its hardships and consequences, overshadowed and dominated the news.

The COVID-19 pandemic split societies along underlying fault lines. Long-standing inequalities and injustices meant disproportionately dire consequences for vulnerable populations. In the U.S., mortality rates among Blacks, Indigenous peoples, and Latinos were far above the average.[1] In fact, taking age into account, Blacks were 2.7 times more likely to die of COVID than were Whites; Latinos, 2.4 times more likely; and Indigenous peoples, 3.3 times more likely.[2] Persons with chronic disease, the elderly, and those in institutionalized settings including prisons and detention centers were also at greater risk.

Like the disease, the economic downturn hit at-risk population groups hardest. Lower-income households were more likely to have a member who lost a job, more likely to have difficulty meeting rent or mortgage payments, and more likely to have relied on food from food banks.[3] When the U.S. unemployment rate hit 14.7% in April of 2020, it left over 23 million persons officially unemployed [this number did not include those not looking for work or those underemployed]. The highest unemployment rates were among Blacks and Latinos.[4] Adding insult to injury, many unemployed workers lost their employment-based healthcare coverage along with their jobs; not until the passage of the American Rescue Plan (ARP) in 2021 could such workers keep their health coverage, if only from April 1 to October 1, 2021 under the ARP.

The closure of schools and child care centers brought increased work for women at home: mothers, more often than fathers, home-schooled their children to supplement online lessons and also cared for their preschoolers. Many women quit paid employment to attend to these duties. Women who kept their job, but worked from home as they supervised children, had double duty, though sometimes shared with a willing spouse.

The change in usual routines and the increased isolation brought a rise in mental health issues, which affected a large percentage of the population.[5] Domestic violence increased. Preliminary numbers show a double-digit percentage increase in domestic violence calls or arrests in the U.S. cities studied.[6]

Although the likelihood of an eventual pandemic had long been acknowledged, countries were unprepared. Neither governments nor the healthcare industry had plans in place covering the basics, despite the warnings of experts. The U.S. had even disbanded its pandemic task force. Severe shortages of hospital beds, personal protective equipment (PPE), ventilators, and more were inevitable.

The U.S. in global perspective. The U.S. fared worse than many other countries in both health and economic outcomes initially. The U.S. mortality rate due to the virus was higher than that in countries including India, Germany, Canada, France, and the U.K., from June through October 2020, as well as at times during the following months. Poor compliance with guidelines of the Center for Disease Control (CDC) in several regions brought soaring mortality rates in the last months of 2021 and into 2022.[7]

The economic downturn in the U.S. began with a sharp drop. The U.S. unemployment rate jumped to 14.7% in April, while in the Euro Area it was a more moderate of 8.4%, due to more generous leave policies.[8] The frayed safety net in the States left many families without even a basic income and often without assured access to affordable health care. Most European countries and Japan, with stronger safety nets, coped better. They were able to crank out needed funds through mechanisms already in place. Workers generally had several months of paid sick leave to rely on as well as payment protection plans, which paid businesses to retain workers. Job-sharing was common and health care was universally accessible.

In contrast, millions of U.S. workers were in dire straits. Although eligibility requirements for unemployment benefits were broadened, workers had a hard time accessing them: state agencies were overwhelmed by demand and had to hire more staff and set up new procedures. Undocumented workers, deemed essential for the nation's agricultural supply, had even greater difficulty accessing any of the benefits provided by the Coronavirus Aid, Relief and Economic Security Act (CARES Act).[9]

The poor numbers for the U.S. were directly tied to such pre-existing conditions as the lack of universal healthcare coverage and inadequate income security programs, but also to a deeply flawed response to the coronavirus. First and foremost, there was no coordinating national policy setting out the steps required to contain the spread of the virus and to ensure the availability of needed resources. The U.S. instead left policy guidance to the individual states and to the CDC. There were severe shortages of personal protective equipment for caregivers in direct contact with patients, shortages of supplies required for testing, and of drugs that were found to be effective in treatment of the virus, like Remdesivir. Ventilators were also in short supply. Just-in-time production in the healthcare industry meant stockpiles were low. No federal agency stepped in to address shortages; no emergency was declared directing firms to produce needed items; and the distribution of existing supplies was left to the mercy of often cut-throat competition among hospitals, firms, and states (price-gouging was, however, prohibited). Fend for yourself was the message received.

Nurses and other frontline workers had to re-use masks for a week or longer due to shortages, although the masks were meant to be changed between patients. One result was a high infection rate among these workers. One study found that healthcare workers were three times more likely to contract the coronavirus than the general population (pre-vaccine), and among healthcare workers of color the rate was higher.[10]

With good reason, then, healthcare workers worried about exposing their families to the virus, but alternative housing was seldom made available to them. Those who could not afford a hotel room were forced to sleep in their cars or find other accommodations on their own. Other workers also deemed essential, whose jobs required extensive contact with the public (such as supermarket clerks) faced the same choices. And quitting the job due to concern about immuno-compromised family members (including themselves) was a costly choice: it could bring loss both of employment and of unemployment benefits.

Measures taken to slow the spread of COVID-19 included quarantining those potentially exposed to the virus, masks, social distancing in public spaces, and at times lockdowns of non-essential businesses and offices. Many people stayed at home. These COVID protocols, especially quarantining, meant that housing became a critical, but under-supplied, need. Large numbers of people left urban areas for rural and suburban ones where physical distancing was easier. Physical distancing also meant fewer beds available for the homeless: shelters were accepting fewer persons in order to keep within safety guidelines. At the same time, demand for hotel rooms plummeted, leaving many empty rooms. These rooms, however, were seldom affordable for those seeking alternative shelter.

Housing was in crisis even before the pandemic. A decline in housing affordability in major cities was a key factor contributing to a housing squeeze on middle- and low-income groups. Over one-third of U.S. households are renters, and about half of these pay more than 30% of their income to rent.[11] [Low-income renters pay an even higher percentage of their income.] A March 2021 report of the Consumer Financial Protection Bureau found that many households were behind in their housing payments: 20% of households renting single-family homes, 10% of households owning single-family homes, and 28% of those in manufactured homes. In fact, over 11 million families were deemed at risk of losing their homes in 2021.[12]

The effects of the virus, then, reverberated throughout the society. The direct blow dealt by the virus impacted the health of thousands, and protective measures necessitated massive shifts in daily routines. The economy tanked. Businesses in retail trade, personal services, travel, and tourism saw sales drop precipitously. Over 130,000 small businesses closed permanently as a result of the pandemic, as well as 70,000 other establishments including, for example, individual businesses belonging to a chain.[13]

Relief packages offer benefits, with bias. The first relief package aimed at addressing the economic downturn and healthcare crisis had a massive price tag: $2.3 trillion. The CARES Act was signed by President Trump in late March 2020. A follow-up package meant to go into effect in the fall did not make it out of Congress, so that the next relief stimulus was the $1.9 trillion American Rescue Plan under President Biden.

The CARES Act focused on economic relief for business. The single largest allocation was $510 billion to permit the Treasury to provide loans of up to $4 trillion (yes, trillion) to big business; another $377 billion went to shore up medium and small businesses, largely through the

Paycheck Protection Program, while $292 billion went to individuals for immediate relief and $262 billion went to boost unemployment payments.[14]

Hospitals, however, remained under-funded. Only $117 billion was allocated to hospitals, many of which were operating under growing deficits. Some loans were available to hospitals—loans extended by Medicare and Medicaid—but these loans became due in September for the most part, an unrealistic deadline. Small hospitals and physicians in rural areas often operate only on subsidies from state governments. Across the country, shortages of PPE and testing kits remained problematic through all of 2020.

Federally backed credit and tax breaks under the CARES Act allowed big business to get through the pandemic largely unscathed. Besides the $510 billion going to the Treasury to enable it to extend low-cost credit to business, the Federal Reserve opted to purchase corporate bonds at its discretion; this decision bolstered investor confidence in bonds issued by struggling companies like Boeing.[15] Boeing, in fact, did not have to borrow government funds as a result of this Fed initiative.

Small- and medium-sized enterprises (SMEs) were allotted the second largest amount, $377 billion, under the CARES Act. Much of this was for the Paycheck Protection Program, which provided funds to businesses with under 500 employees on condition that criteria regarding employee retention and expenditures were met. This program was modeled on similar plans long in effect in Europe. Businesses with fewer than 15,000 employees and less than $5 billion in revenue were eligible for loans under the Main Street Lending Program. Several firms that had more than 500 employees took advantage of the PPP program and many smaller firms received no help as funds quickly ran out.[16] A second round of funding ($320 billion) was authorized for the PPP, which ended in August 2020, and a third and final round was authorized in December.

Besides these direct expenditures, generous tax breaks totaling $258 billion went to upper-income individuals who, it was feared, might lose money in the stock market. As it turned out, though, after an initial drop, the stock market was again soaring by the fall of 2020.

Middle- and low-income households received some assistance. Over $290 billion went to provide all but high-income families $1200 per taxpayer, with an additional $500 per child. Another $260 billion helped the unemployed by boosting unemployment payments by $600 per week and by covering workers who were ineligible for their state's unemployment insurance benefits. [In fact, except during recessions, most unemployed workers in the U.S. do not receive unemployment compensation due to eligibility requirements. Self-employed workers, contractors, workers forced to quit their jobs to care for ill family members, part-time and many part-year workers, and so on are generally not eligible for unemployment compensation.]

The expanded unemployment benefits expired on July 31, and were then continued (at the lower level of $600) from late December 2020 to September 2021. The CARES package, then, left many unable to provide for basic necessities.[17] Even before unemployment benefits expired, lines at food banks were long—and only $25 billion had been allotted to expand the food stamp program. Local communities, then, largely funded the food provided at food banks and by other programs. State and local governments, strapped for cash by a plunge in tax revenues, were unable to meet such basic needs: the $150 billion they received through CARES largely went to cover pandemic-related demands such as the provision of testing and the reconfiguration of government services to prevent the spread of the virus. This later changed.

The CARES Act included a moratorium on the eviction of renters but that expired in August. In September (2020), the CDC imposed a ban on evictions which was repeatedly extended until it was successfully challenged in court and brought to an end on July 31, 2021. Financial assistance for renters was part of the American Rescue Plan of 2021, which had provided funds to states for that purpose. But states were very slow to disburse those funds—most of those funds in fact remained undistributed as of the end of September 2021.

Homeowners with federally backed mortgages also received protection under a CARES program that was extended by the Biden administration (about 75% of all mortgages are federally backed): homeowners could defer mortgage payments for the period covered.[18]

In addition to extending and modifying several provisions of the CARES Act, the American Rescue Plan also included new provisions that expanded benefits for middle- and lower-income families, including the expansion of child tax credits, of Earned Income Tax Credits, and of subsidies for healthcare coverage, as well as greater assistance for housing, for state and local governments and for education.[19]

State administrative offices were swamped by the demands of both the CARES and the ARP Acts. Channels for ensuring that funds reach those qualifying for assistance under new programs had to be hastily set up by individual states. Eligibility had to be confirmed. Application processes were plagued by a lack of personnel and a lack of clear procedures. As a result, unemployment benefits, housing assistance, and the like were often very slow in coming.

Overall, neither the CARES Act nor the American Rescue Plan went much beyond safeguarding the interests of businesses and investors and helping families get by. Despite some improved benefits for families, unemployment compensation dropped under the 2021 relief package. The subsidies that expanded healthcare coverage left the price of insurance high. Households who suffered from job and income loss found it difficult to access assistance that helped them survive. Those who worked in the healthcare sector continued to lack the support needed for their work—despite some improvement, the 4th wave of the pandemic in early fall 2021 (driven by the Delta variant) found hospitals at the breaking point again, still lacking needed resources. Housing remained an issue. The Omicron variant later made the situation even worse.

The development of effective vaccines against COVID-19 in late 2020 was expected to slowly and surely bring an end to the pandemic. While there were glitches and inequities in the initial distribution of the vaccines, by mid-2021 the reluctance of many to be vaccinated had emerged as a major obstacle to ending the pandemic. Misinformation and disinformation (false information that is knowingly spread) fostered distrust in the warnings and recommendations of health and government officials. A militant movement of anti-vaxers emerged among those suspicious of the vaccines, along with widespread anti-mask demonstrations. But a lack of trust in public officials was not new. A special issue of Scientific American (2022) covers the anti-vax protests and other aspects of the pandemic.

The level of trust in the U.S. government had been falling for years and hampered the ability of officials to counter the mis- and disinformation.[20] A low vaccination rate among parts of the population plagued efforts to contain the virus. Deep divisions within the country also contributed to the mistrust of officials, and vaccines. The political dynamics that exacerbate divisions in the country is one of the problems explored throughout this book. [On the other hand, the high levels of trust in public officials have been cited as a reason for the success of countries like New Zealand in combatting the virus.] But trust in public officials is not confined to democracies; China is among the more autocratic governments that maintains a high level of trust.[21]

Concluding Remarks on the Pandemic

One unmistakable conclusion that can be drawn from the health and economic crises of the 2020–2022 pandemic is that the suffering caused was concentrated among low- and middle-income families and individuals. The failure of authorities to address either the critical shortages in health-related areas or people's lack of access to basics like food and housing is made clear by the composition of the two major relief packages. The first prioritized protecting larger firms and upper-income individuals. The second maintained the provisions protecting the well-to-do (adding to the growing wealth gap), although it also provided more funding for various assistance programs, for education, and for state and local governments. The overall effect of both was to protect big business and to enable struggling households to get by, while leaving the patchwork of programs aimed at protecting the general population the same as ever—except in a few areas related to managing pandemics. Even universal healthcare coverage did not emerge as a new national priority.

In the midst of all the hardship, a few did extremely well. The stock market soared past the all-time highs reached in late 2019. In addition, the crisis opened new opportunities for financial gain for those who were well-positioned. The rise in online sales benefited companies like Amazon, UPS, and FedEx. A movement away from cities drove down real estate prices and rents in urban areas, while pushing them up elsewhere. As people were forced to leave their homes, some landlords and investors profited. As we will see, similar activity in real estate also happened in the aftermath of Hurricane Katrina in the Gulf Coast (as well as after other hurricanes).

One result of economic trends during the pandemic, then, has been a widening of the income gap. The federal minimum wage was still at $7.25 an hour in mid-2022. Although there were great differences in the relief policies of the Biden and Trump administrations, high levels of respect for market forces characterized both. Both pursued policies meant to maintain a capable and hard-working labor force, while not increasing wage levels "too rapidly." The debt incurred to finance the relief packages may mean higher taxes at some point, but it remains to be seen how much of the tax increase will fall on the wealthy.

Democratic initiatives to address long-standing problems such as economic insecurity, inadequate health care, unequal access to higher education, and crumbling infrastructure received mixed support. The trillion-dollar bill addressing physical infrastructure proposed under Biden passed with a few Republicans signing on. The other major bill, addressing social infrastructure (Build Back Better), did not make it through Congress, although some parts of it likely will. While such bills are a move in the right direction, decades of neglect of critical long-term issues means that decades of investment will be required to remedy the problems.

New Orleans and Hurricane Katrina

Hurricane Katrina hit New Orleans on August 29, 2005. The levees protecting the city were breached and floodwaters rushed in, gradually inundating the area. An estimated 70,000 persons were left stranded at the Superdome, the Convention Center, and elsewhere in the flooded downtown, with very little water or food. For days, Americans watched in horror as survivors waved frantically, and futilely, at television cameras in helicopters overhead. Relief supplies did not arrive in the city until evacuation began four days later. The situation at the Convention Center and on the nearby highway overpass was even worse than at the Superdome: over 25,000 persons clustered together at those sites without any initial supplies and without any National Guard troops. Eventually, Louisiana counted 1464 dead, with another 135 persons missing—many of them presumably vanished in the floodwaters. Seventy percent of occupied housing in the city was damaged or destroyed, hundreds of thousands of persons relocated, and thousands of businesses bankrupted.[22]

The Federal Emergency Management Agency (FEMA), the lead agency in the response and recovery efforts, was quickly declared a disaster itself. FEMA proved incapable of taking even the basic, critical steps needed to save lives. It was not up to the task of providing clean water, let alone coordinating the many agencies and groups involved; in many instances, it prevented experienced organizations, including the Red Cross, from doing urgently needed relief work. Investigative reports on the aftermath of Katrina by the Congress, the White House, and independent sources eventually documented the many shocking failures.[23]

But FEMA was far from being the only organization that failed the people of New Orleans. The City of New Orleans, the State of Louisiana, the Army Corps of Engineers, the Department of Homeland Security and, ultimately, the White House all came up catastrophically short.

Before the storm hit, preparations by the city were woefully inadequate—in terms of both its evacuation plan and its provision for those left behind. Post-storm, the city was totally overwhelmed. Officials at the state and national levels in turn failed to recognize the urgency of needs that were beyond the capacity of the city to handle. In fact, officials at all levels of government in their first press releases seldom addressed the grim situation of the survivors left in the city but rather focused on threats to law and order, threats that were greatly exaggerated.[24] The thousands of military personnel and police that were quickly dispatched to the city aided in rescuing people from floodwaters and stood guard at outposts to maintain order, but they did not have access to the supplies of water and food that were desperately needed. They were unable to evacuate those stranded for several days and themselves lacked adequate provisions.

A straightforward chronicle of events leading up to and following Katrina-related tragedies does little to clarify why things went so wrong. Very different interpretations of events have been offered. Some observers have reached the conclusion that no one was at fault, while others point the finger of blame at the White House, the City, the State, or even the inhabitants of New Orleans themselves. There are a wide range of targets. Not surprisingly, then, very different measures have been proposed to prevent the recurrence of a similar catastrophe.

An examination of the critical decisions made during the Katrina crisis shows that mistaken priorities guided those leading the response to the hurricane. Even now, the priorities of responders as set by a few persons at the top continue to be controversial, as in the case of Hurricane Maria in Puerto Rico in 2017. News stories document that the response to disasters continues to be mismanaged, with funds misdirected and those affected having little voice in the rebuilding of their communities.[25] Oversight of relief and rebuilding is spotty, and accountability to the public is lacking. Below, we provide a brief narrative of storm-related events in New Orleans in 2005, with a focus on key groups and their differing roles in the unfolding saga.

The Big One Is Coming

A series of articles in New Orleans' largest newspaper, the *Times-Picayune*, in June 2002 warned of the inevitable catastrophic storm, the Big One that would strand at least 100,000 people and kill thousands.[26] It detailed the shortcomings of the city's evacuation plans and exposed the inadequacy of the flood protection system. Perhaps it failed to envision the horrific conditions that materialized in the city, but all told, the report proved to be very much on target. It made an impact when it was published; plans were debated and requests made to Congress, but in the end the system scarcely budged. The state's request for more federal funds for flood protection was denied and, in fact, funding was cut.[27]

Still, the City knew what it could face, yet did little to update its plans. As warnings of the coming storm hit the airwaves in late August 2005, New Orleans began to put its outdated hurricane evacuation plans into effect. On Saturday, August 27, the order to evacuate was issued. One of the few innovations put into place after the 2002 *Times-Picayune* report was successful—the stream of vehicles out of the city was expedited by a redirection of traffic flow on some major routes, so that all lanes carried vehicles away from the city.

Those without cars, however, were largely out of luck—the City under Mayor Nagin decided not to provide transportation to emergency shelters outside the city.[28] Instead, the City collected people left behind and dropped them off at the Superdome and the Convention Center. It left bottled water and emergency rations sufficient for 15,000 persons for three days at the Superdome but nothing at the Convention Center. As it turned out, over 30,000 sought refuge at the Superdome and over 20,000 at the Convention Center.[29] Most were there for five days. There were 550 National Guards at the Superdome, but no troops at the Convention Center.[30] After the power outages that hit along with the storm, the heat inside both facilities became sweltering (the Department of Homeland Security was unaware of the loss of electricity even two days later). There was no running water.

The Coast Guard rescued thousands from rooftops and buildings after the storm. But, after the rescue, many were dropped off on high ground in the city where neither water nor food was available.[31] The provision of food and water was not their assigned responsibility.

Prisoners at the Orleans Parish jail stayed for days in cells that gradually filled with sewage-tainted water up to their chests before being taken to a nearby overpass, still without food or water. A first-hand account of days spent on the overpass near the jail by a stranded book publisher makes sobering reading.[32] Many critically ill patients at Memorial Medical Center hospital suffered as well. Some died. The backup generator there failed early Wednesday, two days after the storm, and medical staff worked out a triage system. Some patients were eventually given what were lethal doses of morphine when it appeared they wouldn't be able to survive the hours of stress of a very prolonged and difficult evacuation process, up or down many flights of stairs in a sweltering building without IVs or ventilators. One well-regarded doctor and two nurses with a leading role in the triage system of the evacuation were later arrested and charged with second-degree murder, but the grand jury did not indict.[33]

Administrative Chaos: Misplaced Priorities or Depraved Indifference?

Reports later issued by Congress and the White House detail the administrative chaos among emergency responders after the hurricane hit.[34] Cell towers were knocked out and land lines were not functioning; satellite phone connections were only spotty. Leaders were forced to rely on isolated reports and rumors, which, in turn, contributed to confusion and mistakes.

In the midst of this chaos, politicians focused their comments on the need for law and order. At press conferences, they attacked looters. But, as the days went by, photos of "looters" heading away from convenience stores with bottles of water and food began to evoke sympathy as viewers became aware that the very survival of those "looters" was at risk.

Still, on Thursday, September 1, four days after the storm and before provisions had reached survivors, the White House press secretary was asked to comment on people who were breaking into stores for water and food. He referenced the President's zero tolerance policy toward looting and said those stranded should continue to wait for relief. "Looting is not the way for them to (get help)," he declared.[35]

The same day, the Governor of Louisiana, Kathleen Blanco, warned that National Guard troops "have M-16s and they're locked and loaded. These troops know how to shoot and kill, and they are more than willing to do so if necessary, and I expect they will."[36]

Why the focus on law and order, and not on getting water and food to survivors and evacuating them from the city? One explanation is that agency heads stuck to their areas of responsibility. If the provision of food and water wasn't part of their job assignment, they did not concern themselves with the problem. But reports from those at the scene, including news reports, made it clear that water and food were desperately needed.

Officials apparently were only too willing to assume that someone, somewhere must be on the verge of providing the needed water and food. The Louisiana DHS even turned down the Red Cross request to deliver water and provisions three days after the storm, saying they needed time to set up distribution centers.[37]

The concerns about law and order voiced at the White House and the State House were echoed by neighboring towns. The city of Gretna, bordering New Orleans to the south, blocked vehicular access even before the storm hit. "According to our disaster response plan, the Gretna Police Department sealed our city limits and prevented anyone from entering for obvious reasons of safety and security," stated the mayor.[38] On September 1, the city began turning back at gunpoint for those fleeing New Orleans on foot—with police firing shots over the heads of those on the bridge to Gretna, including at least one Gretna resident.[39] The fact that the majority of those stranded in New Orleans were Black is often cited as a reason for the poor treatment of the flood victims.

While rumors of violence, especially at the Superdome, were widespread, few instances of assault, rape, or murder could later be verified. "All the talk on the news about rape and murder and assaults at the Superdome – it was all wrong. In reality, the Superdome was a disgusting, filthy mess, and it seemed to always be one lit match away from turning into an inferno. But the fire never started."[40] An army report later found that six people died at the Superdome during the aftermath of Katrina, but none were crime victims.[41]

Adding to anxiety and unrest in the city was the occasional sound of gunshots. Residents stranded in their homes, it turned out, were firing gunshots to attract the attention of rescuers—a practice that, not surprisingly, was often misinterpreted. Rescuers worried by shots they heard nearby often withdrew, prolonging the rescue process and bringing law enforcement to the scene with expectations of hostile gunfire.

In fact, the most infamous incidence of violence during this period began with gunshots of uncertain origin. On September 4, out-of-uniform police responding to a report of gunfire near Danziger Bridge shot several members of an African American family out in search of groceries, killing two. Later investigation found that residents trapped in their homes nearby had fired their weapons in an attempt to attract the attention of rescuers.[42] The focus on possible criminal activity drew the spotlight away from the lack of necessities and the problems of communication which affected early relief efforts.

Flooding Ends but Not the Crisis

FEMA, as the lead agency in the disaster response, became the target of harsh criticism. Besides its inadequate action on critical fronts, FEMA repeatedly made it difficult for others to provide assistance. For example, FEMA prevented the Red Cross from delivering food supplies, turned down the use of a Navy hospital ship, and turned away Wal-Mart trucks with supplies. It issued orders that first responders were not to go where needed unless dispatched through appropriate channels. FEMA routinely ignored urgent requests from local authorities if proper paperwork hadn't been filed (a complaint echoed by the Mayor of San Juan, Puerto Rico, after Hurricane Maria in 2017). FEMA cut Jefferson Parish's emergency communications line, after which the sheriff posted armed guards to protect the (repaired) lines from FEMA.[43] Federal leadership was out of touch with the on-the-ground realities. One participant in briefings on the federal response at the White House reported a "surreal" atmosphere there, with "self-congratulatory" statements being exchanged among officials. The infamous quote from President Bush gives an idea of the mood during those first days. "Brownie, you're doing a heck of a job!" President George W. Bush said to Michael Brown, head of FEMA.[44]

The evacuation of the Superdome and other locations began on Friday September 2 and was largely completed within two days. Those evacuated were sent to shelters, often in other states, including to the Houston Astrodome. Many of the evacuees never returned to the city, in part because little assistance was provided for them to do so.[45] The population of the City of New Orleans proper shrank from 455,180 in July 2005 to 208,548 a year later, more than a 50% drop.[46] Even in 2007, the population remained 35% below 2005 levels despite returnees.

Federal Funds Mishandled

Residents returning to the city in the first weeks after the storm came back to a devastated city. Most could salvage only a few of their possessions: the interiors of their homes had become soggy messes after sweltering in floodwaters for weeks. Those in public housing could not even peruse their belongings: unlike other residents, they were denied access to their homes prior to their demolition.[47] Electricity remained out in many parts of the city for months, affecting traffic and street lights as well as homes and businesses.

Times of tragedy bring out the best and the worst in people. Survivors were aided in the rebuilding of their city by the generosity of many. Countries around the world offered assistance in relief and recovery efforts. Thousands of volunteers from the U.S. and abroad came to help.

But at the same time, the unscrupulous took advantage of the turmoil and disruption, and of the funds flowing in to aid the city. Besides looting, racist attacks, and cases of outright fraud, contractors overcharged, under-performed, and under-paid their workers. Locals with power pushed through pet programs that would have met with strong resistance at other times. Insurance companies mobilized quickly to make initial payments, but attached conditions that made it difficult for homeowners to receive the full payment due under policy terms. [The insurance industry, by the way, managed to achieve record profits that year.[48]] Louisiana's Attorney General filed a lawsuit claiming bad faith practices against several insurers.[49]

Congress allocated an estimated $120 billion in federal funds for Gulf recovery efforts, with about $78 billion of this total going to Louisiana.[50] These inflows stimulated an economic boom for those in reconstruction-related work.

As part of the federal aid package, Congress also lifted many federal regulations. Protections for workers, such as those under the Davis-Bacon Act, were suspended.[51] Sanctions against hiring undocumented workers were lifted.[52] The requirement that a certain percentage of Housing and Urban Development (HUD) funds go to low- and moderate-income housing was suspended. Generous tax breaks were granted to firms located in Gulf Opportunity (GO) zones, areas that could be a hundred miles or more from the region slammed by the storm.

FEMA was the largest single recipient and distributor of federal funds, receiving $60 billion initially. It awarded contracts for debris removal, shelters, policing, infrastructure repair, and other needs. There were severe problems with both the process followed in awarding these funds and in the resulting outcomes.

The positive impact on the city was reduced not only by the suspension of labor regulations but also by contract terms that left the hiring of local workers and businesses to the whim of out-of-state contractors. Congress did not require FEMA to give preference to local firms as available, nor were contractors required to give preference in hiring to locals.

As a result, local firms and workers were at the mercy of trickle-down effects, receiving only jobs and pay as decided by large contractors.

Migrants flocked to the coast—about a quarter of them undocumented by one estimate (Klein 2015). They provided much of the labor under conditions that were exploitative.[53] Pay was low and, in many cases, undocumented immigrants were not paid all they were owed. Several lawsuits were filed on their behalf, with some partially successful.[54]

Although the city's tax base had been wiped out, relief funds could not be used to pay the salaries of city workers—thousands were laid off and work given to others. Among those laid off were many of the city's planning staff, while millions of public dollars went to outside consultants, who were often real estate developers (Klein 2015).

Those receiving the largest contracts sub-contracted out the on-the-ground work. Sub-contractors in turn sub-contracted out jobs, until those doing the actual work at the bottom of the line were left with low pay under unprotected working conditions. Local firms made little, if any, profit. One small firm providing debris removal services in Mississippi complained that they actually lost money on their sub-contract, and stated that they were not alone.[55] Contract provisions even made it difficult for volunteers to assist in relief efforts. For example, local morticians who volunteered to help were forbidden to handle the dead, on grounds that this would be an encroachment on the contract awarded to the out-of-state firm Kenyon. [Kenyon charged the state, on average, $12,500 a victim and, according to complaints, failed to properly label many bodies (Klein 2015).]

Several large firms received contracts without going through a bidding process: Klein lists half a dozen that, taken together, were awarded over $3 billion without open bidding. "Halliburton's KBR unit had a $60 million gig to reconstruct military bases along the coast. Blackwater was hired to protect FEMA employees from looters. Parsons, infamous for its sloppy Iraq work, was brought in for a major bridge construction project in Mississippi. Fluor, Shaw, Bechtel, CH2M Hill—all top contractors in Iraq—were hired by the government to provide mobile homes to evacuees just ten days after the levees broke."[56] It can be pointed out that politicians often had links to some of these firms. Vice-President Dick Cheney, for example, was CEO of Halliburton from 1995 to 2000.

Envisioning a Future

In New Orleans and on the Mississippi coast, meetings to plan rebuilding after the storm were packed. Charrettes led by professional planners and architects generated excited discussion as locals explored new possibilities. Many residents, however, were skeptical that their needs would get priority. Emotional outbursts marked townhall-style meetings. The high stakes increased the tensions and conflicts that were already present in the city. On the one hand, most residents and grassroots organizations shared the vision of a secure and prosperous future that included decent housing for residents whose homes had been damaged; protection from future flooding; the creation of more living-wage jobs; good public schools and public transportation; and access to health care. But, on the other hand, those who ultimately made the final decisions had different priorities, as evidenced by the actual allocation of expenditures for reconstruction and by decisions and events documented in depth by Gotham and Greenberg (2014).[57]

New Orleans, in the end, was reshaped with an eye to building a higher-income locale. Those dominating decisions about reconstruction pushed for and got the revamping and reduction of public housing, a new school system based on charter schools with no teachers' union, and an end to the public Charity Hospital, which had been a primary source of care for those in the center of the city. Within a few years, the entire state public hospital system was privatized. Public transportation received little funding and, rather than improving, has worsened in the post-Katrina years.[58]

Not only were the needs of lower-income residents neglected, their situation in many ways became worse than it was pre-Katrina. Poverty rates rose to 27% in 2016, despite the loss of population. In 2005, pre-Katrina, and again in 2010, the poverty rate was 24%; since 2011 it has been above 26.5%.[59] The poor have been hurt by reductions in low-cost housing, public health care, and transportation. Many low-income residents saw their homes demolished or gutted after Katrina, with their belongings lost. They lacked the funds needed to cover the cost of repair or replacement. Their social networks were in shambles and their access to health care dropped with the loss of hospital beds and doctors. The public transportation network worsened post-Katrina, with the exception of transportation for evacuation of the city.

One study has shown that, in the year after the storm, the mortality rate of the population in New Orleans rose 47% to 91 per 100,000.[60] The closing of Charity Hospital in the city's center contributed to this figure. Funds available for rebuilding Charity were channeled instead into a new medical center designed to attract higher-income clientele.[61]

Public housing serving 5148 families was demolished in the name of urban renewal, and replaced with fewer units and in locations with little or no public transportation. In many cases, the cost of repairing those units would have been less than the cost of demolition. The St. Thomas complex, where redevelopment began even before the storm, had had 1510 units; its replacement, River Garden, has 182. The St. Bernard complex had 1464 units but Columbia Parc has only 221.[62]

Local groups pushed back against large parts of these initiatives, but with very limited success. In one success story, the Neighborhood Partnerships Network, a grassroots effort, mobilized rapidly in early 2006 and managed to block a proposal put forth by Mayor Nagin's commission that would have prevented people in several low-lying areas from rebuilding, without offering them any realistic alternative (Gotham and Greenberg 2014: 98–109).

Flood Protection, and Hurricane Preparedness, Remain Inadequate

New Orleans still remains vulnerable to catastrophic flooding, despite the outrage at the inundation of the city and the expenditure of billions on the levee system after 2005. Progress in revamping the levee system has been slow, and the loss of wetlands remains high at about 16 square miles per year.[63]

Besides effective flood protection, hurricane preparedness involves the capacity for more thorough evacuations, the provision of shelters, and modifications to make housing more adapted to hurricanes and flooding. All are costly. But cost alone does not explain why progress has been so slow. Sufficient funding could be available to cover such expenses with appropriate changes in the federal and local tax codes. Louisiana has a potentially large tax base that could be tapped: it is home to half of the country's oil refineries; its pipelines bear 30% of the nation's total oil and gas supply. The port of New Orleans is a vital distribution point for most states.[64] In addition, New Orleans, with its unique culture and beauty, draws millions of visitors annually, generating billions of dollars.

But policymakers continue to tackle the environmental problems the city faces with a hesitant approach that requires fewer investment dollars, thus ensuring an on-going threat of disaster.

Winners and Losers

Looking at who won and who lost during the Hurricane Katrina epoch tells us a lot about who had power and how they used it. Companies gaining big federal contracts were winners, for example, as were real estate speculators who bought up land at bargain rates, and the developers, politicians, and businesspeople who were in a position to shape the remaking of the city.

But most citizens of New Orleans were hurt by the storm and enjoyed few offsetting gains. Besides coping with the destruction of property and the disruption to their lives, they faced higher rents, higher insurance prices, and electricity costs as well as a decline in public services. American taxpayers, too, suffered as their hard-earned dollars were misused, with federal funds lining corporate pockets while leaving residents in distress.

The poor, including those who were unable to return to the city, were hit the hardest. Residents who were Black as well as poor had their situation exacerbated by racism. Articles in *The Nation* and *Slate* tell tales of overt and covert violence against Blacks during the disaster; and studies have linked the slow response to those stranded to racism.[65] The Black Lives Matter movement began as part of the response to the racism apparent during the Hurricane Katrina tragedy.[66]

Events in New Orleans follow a pattern found after other crises. Naomi Klein (2008) has written extensively on capitalism during disasters, analyzing the strategies of firms who benefit from government contracts, lower taxes, and weakened regulation.[67] She focused on New Orleans in her 2015 article quoted above. Gotham and Greenberg (2014) analyze and compare the reconstruction work done in post-Katrina New Orleans and in post-9/11 New York City; they emphasize that the plans put into effect had been in development for years before these disasters happened. Both Klein and Gotham and Greenberg cite the economist Milton Friedman, who back in the 1970s advocated seizing the moment offered by crises. Crises, he argued, make the politically impossible possible, and offer the opportunity to dismantle programs of the welfare state and to move toward what he termed free market capitalism (Klein 2008: 7).

More recently, in the aftermaths of Hurricanes Harvey, Irma, and Maria, similar patterns are being repeated. Large firms receive lucrative contracts, gaining millions in profits, while homeowners face formidable obstacles to obtaining access to funds and other support they need to get on with their lives.[68]

Concluding Remarks on Katrina

The catastrophic events that followed in the wake of Hurricane Katrina should never have happened. But they did. For the millions of people affected by these or similar events, it is no longer possible to believe that government is at the ready to meet the urgent needs of citizens in times of crisis. As we will see, the story of the 2007–2009 financial crisis reveals a similar failure of responsibility, although in that case the failure was largely one of business. In both cases, those with power misused their authority. They failed to ensure the well-being of those who relied upon them. The institutions they headed lacked, and continue to lack, a system of accountability that penalizes behavior that causes harm, no matter how egregious the wrong committed.

The Financial Crisis of 2007–2009

While news coverage of the financial crisis didn't provide viewers with footage as dramatic as that of Hurricane Katrina, the number of lives lost or destroyed was undoubtedly even higher. Millions of people around the globe survive at the margin of subsistence; their very lives are at risk during the cascade of calamities that are unleashed by a severe economic downturn. And the recession of 2007–2009 was the worst in over 60 years.

The collapse in value of financial assets in the fall of 2008 bankrupted firms, demolished savings and pension plans, and sent cities and countries into economic shock. Public and personal debt levels soared. A spike in unemployment rates and drastic budget cuts followed, spreading devastating hardship. For many, access to basics, from food and shelter to health care, dropped dramatically, stressing family networks and the broader society. Globally, malnourishment rose an estimated 9%, driving up mortality rates.[69] Millions of people lost their homes, health care, or savings, while at the same time public services were reduced.[70] Domestic violence, suicides, and mental illness rose.[71] The final toll of the crisis is not clear.

Numbers in Tables 2.1 and 2.2 hint at the dramatic developments that rocked households, businesses, and governments at all levels across the U.S.

The fall of 2009 saw the national unemployment rate rise to 10% for the first time in over 25 years. The unemployment rates for Blacks and youth were much higher: in December 2009, 48% of Black youths in the workforce were unemployed.[72] Many families were directly impacted. About 35% of workers reported that someone in their extended family

Table 2.1 Economic impact of the financial crisis of 2007–2009

	Trends in key indicators			
	2007	*2010*	*2013*	*2016*
Unemployment rates (%)				
Overall	4.6	9.6	7.4	4.9
16–19 years of age	15.7	25.8	22.8	15.6
Black	8.3	16.0	13.1	8.4
Unemployed for 27 weeks				
Or more (thousands)	1243	6414	4310	2003
Number persons in SNAP	26,316	40,302	47,636	44,220

Sources Federal Reserve at St. Louis for unemployment data; U.S. Department of Agriculture for persons in SNAP

Table 2.2 Trends in income and net worth, 2007–2019

	2007	*2010*	*2013*	*2016*	*2019*
Median family net worth (thousands of 2019$)	$149.4	$90.7	$89.4	$103.5	$121.8
Median family net worth, top 10% (thousands of 2019$)	$1383.7	$1402.1	$1245.9	$1743.8	$1589.3
Median family income (before tax) (thousands of 2019$)	$58.4	$53.9	$51.3	$56.0	$59.0
Share of total income					
Top 1%	21%	17%	20%	24%	...
Top 20%	61%	59%	61%	63%	...
Bottom 60%	21%	22%	21%	19%	...

Source Federal Reserve, *Survey of Consumer Finances*. Data on top 1%, Saez and Zucman (2020)

had been laid off at some point during the recession, and the duration of unemployment was the longest in over 60 years.[73] Median household incomes fell (median must be distinguished from the average) and median household wealth sank 40%—most households lost 25% or more of their wealth.[74] Another shocking research finding was that about 39% of households either had negative equity in their home, had suffered unemployment, or had been behind in their house payments between 2008 and 2010 (Hurd and Rohwedder 2010: 1). The number of persons in the Supplemental Nutrition Assistance Program (SNAP) rose dramatically.

The recession set the stage for a further increase in inequality. Despite the huge drop in asset values and the high unemployment levels, the top 10% of households enjoyed a double-digit growth rate of their real incomes over the 2007–2016 period, while at the same time the real incomes of the bottom 90% dropped.[75] The share of the top 10% in total income rose from 47% in 2006 to 50% in 2015 (Wolff 2021: 48).

The shares of net worth held by the top 10% and by the top 1% also rose. By 2019, the top 10% held 78% of total net worth and the top 1, 38%. The share of the bottom 60%, though, sank to 2.5%.[76] The median net worth of this group was a modest $43,800 in 2019 (Federal Reserve 2020), while the average net worth of the top 1% stood at $27.6 million (Wolff 2021: 50).[77]

Enrollment in the Supplemental Nutrition Assistance Program (food stamps), shown in Table 2.1, skyrocketed by 81%. Foreclosure rates tripled, forcing millions of families out of their homes: from July 2007 through January 2014, about six million homes were sold at foreclosure sales.[78]

National data on evictions are not available for the 2005–2010 period, but statistics from cities like Milwaukee WI, Portland OR, and Allentown PA show the trend. It is estimated that one in eight renters was either evicted or forced to move due to inability to pay over the 2009–2011 period in Milwaukee, and eviction rates there were lower than in many cities.[79]

Government deficits ballooned during the crisis. The federal government spent hundreds of billions to bail out the banks, and hundreds of billions more to pull the economy out of its tailspin. State and local governments increased spending on social services, while tax revenues across the nation plummeted and the value of public pension funds sank.

Greed, Competition, and Crisis

The near-meltdown of the global financial system began when the value of loans, in the form of mortgages and bundles of mortgages, collapsed. Mortgages, as it turned out, had been routinely approved for persons who were financially unqualified.[80] The banks issuing these mortgages could afford to do so because they could bundle and sell their mortgages to purchasers in secondary markets. These purchasers used formulas to guesstimate the risk they were assuming. They guessed wrong.

Prior to the crisis, the biggest banks had assumed excessive risk, not only in the mortgage market, but also in other forms of collateralized debt obligations (CDOs) and in even more esoteric financial instruments, such as the now-infamous credit default swaps (CDSs).[81] The Financial Services Modernization Act of 1999 had opened these riskier investments to banks; before then, banks were not permitted to buy and sell stock or private sector bonds, let alone trade in CDOs, CDSs, or other such products and derivatives. Those investment domains were reserved for investment firms such as Bear Stearns and Lehman Brothers. But banks found the lure of the high-profit, risky investments irresistible; they lobbied for, and in 1999 got, access to those markets—arguing that the increased competition would benefit consumers.

Among the many risky financial products, credit derivatives including CDSs played an especially significant role. The U.S. Treasury calculated the risk exposure in financial derivatives of the 25 largest U.S. banks to be $1.3 trillion in the critical third quarter of 2008.[82] The total value of those derivatives was $175.4 trillion. In comparison, the entire value of all goods and services produced in the U.S. in 2008 was $14.7 trillion. The fact that credit derivatives are routinely purchased at the margin increases the instability of derivatives markets. Margin purchases by definition have only marginal backing, and in mid-2008 many investors held highly leveraged positions—so that as they were called on to cover the purchase price of their investments in full, investors were forced to either sell assets to meet their obligations or default. A domino effect was put in motion.

Mortgage default rates were soaring as September began, and on September 6 the Federal Housing Finance Authority put Freddie Mac and Fannie Mae into conservatorship. These two government-sponsored enterprises owned or guaranteed half of the U.S. $12 trillion mortgage market at the time. Nine days later, the bankruptcy of Lehman Brothers, with over $600 billion in assets, followed. Lehman, less than a year earlier, had been the fourth largest investment bank in the world.[83]

By mid-September, it became clear that hundreds of billions of dollars of financial assets held by big banks were not only over-valued, but illiquid "junk." Faith in these giants was eroding quickly. The willingness of businesses to accept credit extended by them was dropping and lending threatened to screech to a halt. Markets were in turmoil. Without the credit cards, lines of credit, and other forms of finance backed by the big banks, business as usual would cease. Decisive government action was urgent.

Congress balked at the first bailout package proposed but, in early October, approved the Troubled Asset Relief Program (TARP), authorizing the expenditure of up to $700 billion. Unprecedented steps taken by the U.S. Treasury and the Federal Reserve enabled markets to keep functioning. The price tag of the bailout was high, and the later stimulus package added to the taxpayer burden. The crisis contributed to the ousting of George Bush and the election of Barack Obama to the presidency in November.

The Federal Reserve began by pumping more liquidity into the system.[84] It dropped interest rates on its loans to banks. The Treasury, enabled by Congress, purchased securities issued by troubled banks—at prices much higher than market price. Its preferred stock purchases under the Capital Purchase Program eventually exceeded $200 billion.[85] Despite demands for the temporary takeover of some big banks, including Citibank, that step was quickly ruled out. Bailouts had few conditions attached, beyond the gradual repayment of the heavily subsidized loans.[86] Congress made no provision to ensure that middle- and low-income persons, including homeowners, would benefit from the low-cost funds provided to banks (although Obama had promised otherwise).

As people across the country lost their homes and pensions; as cities and state governments saw their revenues plummet and investment values shrink; and as unemployment rates soared to double-digit levels, those at the top of the financial industry seemed largely unaffected. Bonuses to financial CEOs remained close to the extravagant pre-crisis levels, at over $20 billion in 2009; company parties at lavish resorts continued; and the usual perks of private jets and luxury hotel accommodations went on as before.[87] Editorials across the nation voiced outrage at the huge bonuses paid financial executives, while middle America received no financial relief despite footing the bill for the bailout. Government officials could have required banks to provide low-cost refinancing for mortgages on homes that were underwater; they could have demanded that banks offer lower interest rates on student and other loans to consumers and small businesses. Some Wall Street observers voiced surprise that holders of CDSs, unlike small investors in other products, were paid in full; they noted the need for better treatment of smaller investors.[88] Except for loans repaid and fines collected from some banks, the cost of the bailout fell to taxpayers. Banks receiving loans of publicly controlled funds were not even required to lower interest rates charged to consumers!

In addition, government officials failed to prosecute those whose actions had led to the crisis. To date, no one at the very top of too-big-to-fail banks has been imprisoned in connection with the misconduct and fraud that led to the crisis.[89] And money from fines paid to the U.S. Treasury has not made it into the hands of those who lost their homes or their savings due to banks' mismanagement. The cost of the crisis, then, totaling in the hundreds of billions of dollars, has been borne, not by those responsible, but by others, including by those who have to pay off the government debt incurred to finance the bailout.

Steps Not Taken

Those who came out ahead in the high-stakes gaming that led to the crisis have opposed new regulations that would restrict their risk-taking. The Wall Street Reform and Consumer Protection Act, also known as Dodd-Frank Act of 2010, is a compromise measure worked out in Congress with extensive industry input. It represents Congress's attempt to rein in systemic financial risk. In this, they have failed: financial experts across the board agree that many banks remain too big to fail and would be bailed out should a similar crisis hit in the future. In fact, the risk exposure of the top 25 banks increased in the aftermath of the crisis, as measured by the U.S. Treasury's Comptroller of the Currency. The ratio of credit exposure to capital was 101 in the fall quarter of 2016, higher than the 91 in the same quarter of 2008.[90] In dollars, credit exposure was $1.0 trillion in Q3 of 2016 on $177 trillion in derivatives. Several big banks have gotten bigger as well.

Dodd-Frank requires more transparency in transactions, with the goal of increasing the ability to assess risk; however, significant loopholes in the legislation undermine its effectiveness.[91] There are higher capital and liquidity requirements but, according to some experts, these are not high enough to have prevented the crisis. In addition, regulators have considerable discretion with regard to the specifics of implementation, and have often chosen to postpone change or to offer broad discretion to the banks in their interpretation of requirements.

Credit rating agencies, like Moodys and Standard and Poor, continue to be paid by the firms that they audit. [An alternative suggested was that rating agencies be chosen at random among qualified firms and paid from a fund set up for that purpose, financed by fees on banks.] The Orderly Liquidation Authority requires banks to absorb the cost of bank failures over time, but only after the fact.[92]

Another crisis and bailout seem inevitable. As *Forbes* noted in 2013, "the underlying cause of the recent financial crisis remains a threat."[93]

What have banks learned from this crisis, this market correction? Apart from meeting new requirements, they have made a few other changes—for example, their prices for CDSs are higher and they have been more cautious in granting business loans and in mortgage lending. But, as noted above, they have not substantially lowered their level of risk exposure in derivatives. Business for the most part goes on as usual.

The middle class and the poor continue to recover from the effects of the crisis. The median value of net assets held by Americans remained below 2007 levels as late as 2019, as Table 2.1 shows. And, while unemployment rates dropped below 5% from 2016 to 2019, the real average weekly earnings of workers remained lower than in 1976.[94]

At this point, it is appropriate to ask why banks, and banking officials, have been able to escape stiffer penalties and regulations, let alone jail time. What kind of hold do big banks have over politicians that has enabled them to have so much influence on the terms of bailout, on the new regulations adopted and on the wheels of justice? The answer is clear: their control of a large concentration of wealth gives them tremendous power.[95]

Finance rakes in the profits. The share of total U.S. profits going to financial institutions has been greater than that going to manufacturing since 2001. Manufacturing's share has shrunk to 19%, while the share of the financial sector is up to 28% (2019).[96] The recession caused only a brief blip in the fortunes of finance. The share of finance in employment, in contrast, has hovered at about 4%. The percentage of workers employed in manufacturing is more than double that at 9%.[97] The lure of the high incomes available has fueled the growth of the financial sector, and has diverted efforts away from productive and into speculative activities. The resulting concentration of wealth has added to growing income inequality. The careless approach to risk has led to systemic instability.

Finance is also a highly concentrated industry in terms of ownership. Total banking assets were an impressive $16.4 trillion in 2017 (almost equal to the value of U.S. GDP that year) and the top 10 banks in the U.S. held 75% of those assets.[98] The concentrated power of finance puts it in a position to defend its interests, and it does. In the process, it incidentally shapes the functioning of our economic and political systems.

Post-facto analyses of the 2007–2009 crisis by economists have focused for the most part on the problem of risk. See, for example, the survey of scholarly articles on the crisis in the *Journal of Economic Literature* (2012). The concern of the authors surveyed is improving the understanding of risk and risk exposure. Their focus is narrow and leaves out of consideration the exercise of power in the sector. Key questions regarding who is affected by risky investments and how are left unaddressed. Analysis of the costs and benefits, by group, of the operations of the financial sector is needed. Also of interest, but not studied, are such questions as: who decides what levels of risk are acceptable? How would any new, improved standards be put into place and who would provide effective oversight? All of which leads to a most critical question: in whose interest is the highly concentrated power of finance exercised?

The experience of other countries during this crisis has been different from that of the U.S. Canadian banks, for example, operate under stricter regulations that reduced the risk exposure of banks there and so enabled the country to avoid the financial instability suffered elsewhere. Iceland, on the other hand, allowed its banks to go under. The major banks were temporarily nationalized and, though the economy went into a deep dive, the country enacted policies that enabled a rapid recovery.

Concluding Remarks on the Financial Crisis

If Hurricane Katrina and the pandemic shook our faith in "Government," the financial crisis shook our faith in "The Market." As consumers, we are led to believe that competition for our business brings us low-cost, high-quality products. In the case of finance, such products include mortgages, checking accounts, and credit cards. The greater the competition, the better things will be for us, we are told. But the 1999 financial reforms intended to increase competition and so benefit consumers had the unfortunate, if not unexpected, effect of increasing risk-taking by banks as well.

After reforms lifted restrictions, bank holdings of risky assets soared to dangerously high levels. These risky investments brought double-digit profit rates to banks, but also led to the near-collapse of the banking system in 2008.

The negative effects of the crisis were unequally distributed. The net worth of the top 1% dropped for a while, but the drop experienced by the bottom 80% was even steeper. Over 30% of homeowners owed more on their mortgages than the market value of their homes even in early 2012. Millions of persons lost savings, jobs or homes, their health, or even their lives.[99] Taxpayers paid dearly: trillions of tax dollars in countries around the world went to bailing out banks and other institutions, while governments were largely ineffective in protecting taxpayers, vis-à-vis financial institutions, from shouldering the resulting debt (IMF 2010).[100] At the same time, individuals with the greatest responsibility for the crisis—those controlling large financial institutions—suffered little or no adverse consequences personally for their roles in creating the debacle.

The bulk of income and capital in the economy passes through banks, making a sound financial system with a stable currency an essential public good. At the same time, the current system concentrates enormous power in just a few banks. This makes the task of regulating financial institutions politically difficult. Because of their clout, the financial elite have been able to shape the policies and practices of the sector. The consequence in 2008 was systemic global instability. The laws and regulations that should have protected citizens instead enabled debilitating risk-taking. Control of the financial system has not changed much since 2008. It now seems close to impossible to redirect finance to its proper role—that of supporting economic activity in service of the well-being of all.

Stories of New Orleans, the Financial Crisis, and the Pandemic Expose Systemic Flaws

Our brief overview of the COVID-19 pandemic, the Hurricane Katrina disaster in New Orleans, and the near-meltdown of our largest banks makes it clear that these crises were not due simply to the shortcomings of a few individuals. Rather, the problems were systemic in nature. Those in charge of the response to Hurricane Katrina, to the pandemic, and those overseeing financial operations in our banks acted in line with the priorities of their institutions. They made decisions they thought would win the respect of their peers. People in jobs at every organizational level, from loan officers at banks to police officers on the streets of

Louisiana, followed the directives of their superiors—whose priorities they often shared. In hindsight, it is clear that those priorities were often catastrophically mistaken and a cause of tremendous suffering. Much greater weight should have been placed on people's well-being when decisions were made; we need new institutional practices that will ensure leaders, in fact, serve those they are meant to serve.

The shortcomings of our political and economic systems affect us not only in times of disaster, but daily. Our social structure is sadly lacking: health and dental care are unaffordable for far too many; homelessness is epidemic; childcare is under-provided; and many young adults lack access to college and professional training. Our bridges and transportation infrastructure are in a deplorable state (finally funding is being provided for upkeep). Workers are stressed: the real hourly wage rate has been stagnant for over 40 years and, when there is not a recession, most unemployed workers scrape by without unemployment compensation. Schools are chronically under-funded, while the wealth of the wealthiest, and income inequality, are skyrocketing.

Not Too Concerned

As a group, those who set institutional priorities have not put these issues at the top of their agendas. Rather, other concerns are more pressing to the elite few who dominate business and politics (they are largely in the top 1%). Their implicit cost–benefit analyses of legislative proposals, for example, typically put greater weight on net gains to business than on benefits to consumers and employees. These elite exercise their influence either directly through the positions of authority they themselves hold or indirectly through their impact on the choice of those heading businesses and running the government. What they say carries great weight with policymakers, legislators, executives, and business leaders. Their opinions have clout.

In the business world, top priorities of elites may include protecting U.S. interests in the global economy, ensuring that governmental restraints on business are kept to a minimum, or safeguarding enforcement of laws that protect business transactions and property rights. Politicians, on the other hand, when setting their priorities, balance their economic and social goals for the nation with their personal goals for the future, including funding their campaigns. In many cases, the jobs they hold either before or after their time as lawmakers depend on the approval and goodwill of one or more powerful interest groups.

Leaders in the business world are in charge of determining how profits are spent, what wages are paid, the content of major media offerings, what investments their companies make and how much financial support different political campaigns receive. As we examine further in Chapter 3, their power is sufficient to ensure that they exert great influence at high-level gatherings.

The elite few, in fact, shape and reshape how markets, government, and other institutions function. System-wide characteristics such as political and financial stability are affected by their actions. This is the framework within which our society functions and within which we approach national problems.

Consider how we approach problems such as disaster response, financial regulation, health care, infrastructure, schools. We have what seems a reasonable process: we ask experts to analyze the problem at hand; then we engage in widespread discussion and debate of their recommendations; and finally Congress and state legislatures enact legislation and regulations that are meant to correct existing flaws. As is the case with our political system generally, this process is heavily influenced by the wealthy and powerful. Ask yourself: Who pays the experts studying the issue in question? Which organizations and groups have the greatest influence in determining which proposals are ultimately adopted?

Power: Prevailing and Countervailing

Surely there are processes, both public and private, that would yield better results. To explore how these might be found, we begin with the premise that a basic goal is that business and political leaders put greater priority on serving people's best interests. We also want these top decision-makers to be accountable to those who are affected by their decisions—which often include all in the U.S. and even beyond. Is it possible to convince the current leadership that they should change their priorities? Can we find ways to hold our leaders responsible for the effects of their decisions?

Reform efforts Let's take a brief look at efforts underway to convince business and government leaders of the need to be more socially responsible, and consider why results have been so limited.

Many reform efforts by nonprofits focus on persuading businesses and politicians to adopt more socially responsible agendas or goals. They work to persuade as many groups and firms as possible to pledge their commitment to environmental-social-governance (ESG) goals. Over 7000 firms have committed themselves to sustainability and human rights practices by signing on to the U.N. Global Compact. In addition, groups like the Interfaith Center on Corporate Responsibility have persuaded hundreds of corporations to work with them for greater corporate responsibility. Shareholders belonging to groups like the Interfaith Center are submitting resolutions at annual corporate meetings that are changing business policies.

Besides such efforts by nonprofits, many firms, including cooperatives and benefit corporations, have made commitments on their own initiative to goals such as those found in the U.N. Ten Principles of social responsibility. They believe that firms that do good also do well. In brief, there are many efforts underway to increase corporate responsibility through promoting voluntary commitments by firms. These firms, however, pick and choose which social commitments they make, and there is no penalty if they do not honor their commitments.

Efforts by nonprofits and others that focus on elections and legislation seem to have more potential. An engaged citizenry, after all, has the power to force significant changes in laws and institutions. Beyond supporting specific political candidates and pieces of legislation, people can organize new political parties and, over time, can even put in place new forms of government. There is currently action around hundreds of issues, including campaign reform, health care, social justice, and protection of the environment. Yet to date, results are inadequate.

Transformative change is a long, slow process; and, unlike most reforms, requires a widespread change in thinking and values. In the U.S. at present, there are no signs that transformative change of our economic institutions is on the immediate horizon. We are likely stuck with the current rules of the game for a while. Pushback against specific proposals that would bring major change has been, and will continue to be, strong.

For example, legislation that would require social impact statements from businesses is controversial; similarly, legislation that would impose severe penalties on businesses that repeatedly break the law or defraud customers sparks heated debate about the appropriate role of government and the value of free markets. Long-standing differences of opinion come to the fore, raising such questions as:

Would we lose out in global competition if firms here followed higher ethical standards? Are markets, based on seeking high profits, necessary for a flourishing economy? Which rules regulating the market actually bring change for the better? Does government represent the voice of the people, the only check on big business? Or is government a tool of the wealthy that rigs the rules of the game to favor big business?

A similar series of questions about individual rights and the role of government come up in debates about election reform measures, e.g. measures that would require that only public funding be permitted in election campaigns. Far-reaching institutional change cannot be achieved simply through electing a few new politicians or adopting new party platforms. Rather, it is a slow process during which each side works to strengthen its base of support; opinions shift; and, if a new consensus is reached, one side prevails but with understandings on both sides modified.

Changing the priorities of national leadership and making our leaders accountable for their decisions is not an easy task. History provides insights into the range of challenges and obstacles that would be involved. Conflicts, for one thing, have too often been part of social change. People feel strongly about change that challenges prevailing beliefs and values, particularly when their own social roles are affected. The spread of Christianity, for example, involved the persecution of the early Christians and, later, Christians led the armed conflict of the Crusades. The founding of the U.S. entailed a revolution. Even less significant change often faces hostile, determined opposition. This is the case with laws that change the rules of the game, such as those permitting limited liability corporations.

What Is Power?

Power in its most basic form consists of the ability to act or produce an effect. This definition is only a starting point for understanding the many forms of power. Institutionalized power, that is, power that is backed by institutionalized authority, is a widespread form of power. But there is always opposition to the prevailing powers, at least in some form. This countervailing power can result at times in profound social change. This book focuses on the problem of changing institutionalized power through deliberate collective action. As background, some understanding of institutions and the sources of the power embodied in them is necessary.

The Social Triad of Institutions, Identities, and Knowledge Let's delve into the process of institutional change, using a simplified framework of social relations, based on the social triad of institutions, individual identities, and shared understandings as described by Peterson (2002). The origins and constraints on power of those in positions of authority derive from this triad. A brief introduction to each of the three members of the triad clarifies their close inter-relationship. We build on the discussion below in future chapters.

Institutions embody our collective power As members of society, we know how we are expected to act in many situations, and also how others will react if our behavior is unusual or deviates from the norm. If, for example, we do not shake a hand that is offered during introductions (pre-COVID, that is), we would expect a reaction showing negative surprise. Social norms such as shaking hands or driving on the right side of the road are shared practices. These rules of behavior can be informal or formal. Shared rules and practices make up the institutions that characterize a society. Institutions, then, are forms of cooperative behavior that structure relationships within a social group or groups. They constitute the cement of a society.

Examples of institutions include the family, the firm, the market, private property, and religion as well as the democratic institutions of the free press, trial by jury, and elections. The legal system, the healthcare system, the financial system, government, and even language itself are also institutions. That is, they are, at their core, forms of behavior with known rules and practices. We perpetuate social institutions over time as we behave in accord with these accepted social practices. Our institutions change gradually over time as we change customs and laws. Note that although "the family" is an institution, there are millions of individual families that follow to a greater or lesser extent the relevant institutional norms. The same is true of other institutions—each unit has its individual character.

Institutions limit behavior in many ways but also, and more importantly, enable individuals to achieve goals from the mundane (getting to work safely) to the epic (walking on the moon). Institutions through laws, regulations, and customs enable basic social functions. They guide how we provide goods and services, law and order, childrearing, and more. The power of a society rests to a large extent on the strength of its institutions. Or, put differently, those who accept the norms and authority of an institution and act accordingly create the power that enables social accomplishments.

At the same time, institutions endow considerable power to those who head its individual units and even more to those who head decision-making bodies at the top of a hierarchy of organizations, such as the hierarchy within the judicial system or that within a large corporation.

Individual identities Everyone belongs to several institutions and social groups, and each may affect his or her identity. Individuals identify with social groups linked to such characteristics as gender, class, race, location, religion, or college attended—even if the categories are imperfect or problematic. Each individual also has his or her own knowledge, history, and experiences. All these factors go into creating the unique individual identity that directs a person's behavior and choices.

The individuals in a group in turn impact that group, affecting its characteristics and what it can achieve. Note that not only do institutions impact behavior, individuals impact institutions. In fact, institutions change as a result of behaviors initiated by individuals.

Knowledge and meaning systems While members of a society may have differing opinions about the merits of its various social practices, there is also a shared understanding about what those practices are that enables individuals and groups to work together. This shared knowledge is not static. Any group in society can create or modify the meaning systems that guide its shared practices, adapting its knowledge and beliefs to changed circumstances. The Church of Believers, for example, will adapt its teachings to guide members as they face new situations—perhaps offering its interpretation of what justice or love requires today. Congress has its own understandings, as do the American Association of Retired Persons (AARP) and the National Rifle Association (NRA). But there are commonalities and overlap in the knowledge and values of most groups in a society, enough to enable meaningful communication and debate. As an example, consider the shared knowledge that vaccines can prevent disease, although beliefs about the desirability of getting vaccinated differ among groups.

The shared values of justice, fair play, and democracy are values that unite the country and are important to facilitating acceptance of election results, of jury verdicts, and of umpires' decisions. The shared beliefs and values that shape the meaning of events within a society are critical to social cohesion and community—they are arguably even more critical than shared factual knowledge.

Those with power due to wealth or to the position they hold within a social group hold social power. Social power necessarily rests on shared knowledge of accepted practices within social institutions and is enhanced by shared values and understandings. The critical role that knowledge and culture play in inheriting, accessing, and holding on to power has long been explored by scholars.[101]

Transforming Institutions

Shaking things up is an uncomfortable business, and restructuring institutions does just that. It tosses some accepted ways of doing things out the window. Social innovators, as they try to fix what they see as broken in our institutions, challenge prevailing views of how the world should work. Those who prefer to keep things as they are will react strongly, and negatively, to these change agents. In fact, these opponents of change constitute the main obstacle faced by social innovators.

Still, it must be recognized that there is a sharp contrast between the change that occurs within current institutional structures in a top-down fashion and the kind of change characteristic of grassroots social movements, or bottom-up change. Top-down change, initiated and controlled by those at the highest levels of power, is a constant occurrence. Those in power are continually adapting practices and policies to deal with an ever-changing world; this flexibility protects their positions of authority. On the other hand, those who challenge the practices that uphold the dominance of a powerful group must do so without such institutional support. The women's suffrage, gay rights, and civil rights movements are examples of such grassroots, bottom-up struggles.

World War II provides a striking example of top-down change. In just a few months, national leaders were able to radically transform work activity and household consumption. Millions of people took on new jobs, many consumer items were rationed, and hundreds of thousands of people joined the military and risked their lives on battlefields. This transformation could be accomplished quickly because there was a unity of purpose and because the power of those heading our national institutions was honored.[102] The transformation of the country during WWII shows the tremendous power that is embedded in national institutions of both the public and private sectors. Similarly, after 9/11, governmental power made possible a swift revamping of security protocols at airports and federal buildings across the country, under the supervision of the newly created super-department of Homeland Security.

Power embedded in institutions, then, can effect massive national change rapidly. However, the bottom-up changes that raised the status of women, Blacks, and the LGBTQ+ community or those that increased workers' legal protections, were the result of power exercised by large, motivated coalitions. In these social movements, a subordinate group confronted and prevailed over the dominant, changing laws and customs. Challenging prevailing practices challenges beliefs and values, however. People's identities are tied to their beliefs. More deeply rooted the traditions, the more central they are to beliefs and identity, and the stronger the reaction to change.

The women's suffrage movement illustrates the kind of challenges facing grassroots movements. The movement began with only a few members and had limited financial support. Years of meetings, discussions, communications, and networking were invested in the effort to gain the vote. There were heated confrontations and internal dissension; some protest leaders went to prison.

Emotions run high on both sides of any mass social movement because of the close ties between identities and the values built into existing institutions. Although the suffrage movement focused on the right to vote, it was immediately recognized that if women gained the right to vote, their status would change not only in the political arena but at home and at work. The institutions of family, firm, and government each had a division of labor between men and women. Gender roles were justified by widely held values and attitudes. Many people instinctively worried that if women got the vote, they would question their traditional roles and might fight for higher pay and an end to discriminatory practices that kept them in certain jobs and barred them from others.

The Changing Attitudes Toward Women in the Past Century

Social change is made more difficult by the impact it has on the social status of different groups. Improving the social status of one group can mean that those in other groups feel they have lost status. They are then compelled to change their sense of identity. Consider women. The lower status of women in comparison to that of men has a history dating back hundreds of years. Achieving equal status can be expected, then, to take a very long time. Change that elevates the status of women changes the sense of identity of both women and men. Those who feel their sense of identity is threatened by this change will likely oppose it; this opposition can become a serious political obstacle. Not all social movements will mount such a challenge to prevailing identities, but new laws and practices benefiting a disadvantaged group often bring some adjustments in people's understanding of that group and their own sense of identity.

Social opportunities and self-identities change for groups that advance, but others may feel resentful. In the early 1900s, women who worked to gain the right to vote faced fierce opposition. Many people saw in women's suffrage a threat to social order. After all, for hundreds of years, a woman's role had been to help, even to obey, her husband and to take care of family and home. Her participation in public life was seen as optional at best. A change in women's traditional role, it was feared, would hurt home life and take away from men's (superior) status in both public and private life. Opponents were right in one sense: more than women's ability to vote was at stake. The women's suffrage movement, once successful, not only gained the vote for women, but also revised the roles of both women and men.

New institutionalized practices change shared understandings about how things are done and why. Attitudes toward women changed along with their newly won rights. In 1900, many men believed the interests of women to be perfectly safe in the hands of men, so that women didn't need the right to vote. Many also opposed women playing in orchestras—wind instruments in particular: "Nature never intended the fair sex to become players of wind instruments… women cannot possibly play brass instruments and look pretty," opined one musician in 1904.[103] The entry of women into jobs previously reserved for men during World War II was facilitated by the improvement in women's status that was an effect of the suffrage movement.

The increased respect for women and their abilities brought a heightened awareness of the paternalism and discrimination that women faced. Women became increasingly unwilling to put up with abuse and unfair treatment, whether at home or at work. They became more active in public life. Ideas about women's abilities and what constitutes fair treatment of women continue to evolve.

Our institutions do not yet safeguard women's rights in many areas; much of the reason for this can be related to the advantages men gain from being in a favored position. For many men, their identity is also tied to a sense of male superiority. The complex of laws and customs that shape the treatment of women today evolved over hundreds of years, and it is not a surprise that full equality remains a goal, not an actuality.

The passage of the 19th amendment granting women the right to vote in 1920 marked a major break-through. Such break-throughs take time; it is not easy to change widespread ideas about a large population group and to put in place new ways of doing things.[104] Knowledge and meaning systems, individual identities, and institutionalized practices are all affected by any significant break with past practices.

Social innovators/change agents working for systemic change from the bottom up, then, are presented with a task that seems Herculean. What about the changes needed today to make our democracy better serve us? To restate our end goal: we want the well-being of people and planet to be a high-priority goal of both the public and private sectors. We want government and business to be more accountable to those whose lives are greatly impacted by their decisions (that includes most of us).

The achievement of this goal will require broad grassroots support; those at the top will not be leading efforts for such change. We will explore why this is so in greater detail in coming chapters. For now, we note here that there are trade-offs and costs involved in changing national priorities. Goals such as improving access to workforce training and education, expanding healthcare coverage, reducing carbon emissions, and providing paid parental leave come at a price. The cost could be covered either by business, through lower profits or lower pay for top management, or by government, through new taxes, most likely on the wealthy who, after all, hold the greatest concentrations of wealth and income. Changes in laws and government regulations would also be needed to align with the new goals. One effect of the new priorities is certain to be unpopular with the wealthy elite: changes like these would cut into their ability to decide how they run their businesses and how they use the resources they now control. They will not cede such power unless and until they are confronted with a highly compelling reason.

Effecting change that takes power away from the powerful is always a daunting task. Those dominating societies are formidable opponents, with an array of resources. Because of their wealth and positions, they have extensive legal and financial backing. Their access to the media allows them to publicize their viewpoints widely and to spread messages that divide their opponents.

The legal and judicial systems provide them strong support, through enforcing laws that protect their prerogatives: laws generally protect the prevailing order and typically restrain the efforts of reformers. For example, workplace organizers face severe restrictions on what they can do at the workplace. Renters with exploitative landlords must hire lawyers if they want to sue landlords for needed repairs (cities often do not provide effective protection). Women who complain about sexual harassment at work run the risk of losing their jobs, or being passed over for promotions. In general, subordinate groups must overcome obstacles that are legal, organizational, and financial before they can achieve gains vis-a-vis the more powerful.

Still, time inevitably brings change. New technologies, shifts in population demographics, the depletion of resources, and the spread of new ideas and values are among the catalysts that can spark reform efforts, or revolution. New circumstances give birth to new coalitions willing to work to improve their future.

Here and now, what next? The top-down, non-representative decision-making that pervades our major institutions undermines our ability to grapple with the urgent issues we face. Leaders in both the public and private sectors are not accountable to those whose fates rest on their decisions. They fail to address the root causes of the dire situations that are only too common. Instead, they provide only minor relief. No end is in sight to problems like widespread economic insecurity, growing inequality, an inhumane prison system, inadequate health care, and environmental degradation. Even worse, the continuing rise in global temperatures threatens the very existence of much of earth's flora and fauna. A shift in priorities toward the well-being of people and planet rests on a shift in power toward those who are devoted to the common good. In brief, we need to build the countervailing power crucial to strengthening our democracy if we are to effectively take on the challenges confronting us.

What do we know about how to create more democratic institutions? Historically, social movements such as the labor movement, the women's movement, and the civil rights movement have been critical to advances in democratic rights in the U.S. We can also learn from movements in Europe and elsewhere that have achieved tremendous gains for their citizens. Social movements and activist efforts have always been vital to advances leading to fuller and more fulfilling lives.

While few would speak against strengthening democracy in our country, in practice there is no simple way to make that happen. We are told that improvements must come gradually through working within current institutions. We are told that to accept the status quo is to be realistic—things are what they are. Our policy analysts are content to ask the following question in varying forms: what new laws and policies will improve outcomes within our system as it now stands? Only a minority of scholars, along with many activists, argue for creating new, more democratic structures. Research that looks for ways in which to achieve a democratic transformation of our institutions is indispensable to the success of such efforts. There are critical questions that need to be addressed.

We are raising some of those questions here: Do the actions of our ruling elite, overall, influence outcomes to their benefit and to the disadvantage of the rest of us? How can our institutions be structured so that the economy, the government, and our leaders, better serve us? Can a government that is of, by, and for the people rest on an economy based on businesses that operate of, by, and for themselves?

We must also ask questions related to strategies for reshaping our institutions: How should we go about changing the current situation? Who can organize and consolidate the countervailing power that can bring changed leadership? In the coming chapters, we take up these questions as we examine power, politics, and the economy.

Notes

1. https://www.apmresearchlab.org/covid/deaths-by-race#age. As of March 2, 2021, the age-adjusted rates were Indigenous 256/100,000; Black 124/100,000; Pacific Islander 90/100,000; Latinos 87/100,000; White 76/100,000; Asians 52/100,000.
2. https://www.apmresearchlab.org/covid/deaths-by-race#age gives the results of a March 2021 analysis by the APM Research Lab. Asians had a below-average COVID mortality rate.
3. https://www.pewresearch.org/social-trends/2020/09/24/economic-fallout-from-covid-19-continues-to-hit-lower-income-americans-the-hardest/.
4. https://www.bls.gov/charts/employment-situation/civilian-unemployment-rate.htm. Data also show that the unemployment rate among women was above that of men during the first months.
5. https://www.kff.org/coronavirus-covid-19/issue-brief/the-implications-of-covid-19-for-mental-health-and--use/, https://www.nimh.nih.gov/health/statistics/suicide#part_2557, and https://bjs.ojp.gov/content/pub/pdf/ndv0312.pdf.

6. https://www.ncbi.nlm.nih.gov/pmc/articles/PMC7195322/. "Alarming Trends in Domestic Violence". https://theconversation.com/domestic-violence-calls-for-help-increased-during-the-pandemic-but-the-answers-havent-gotten-any-easier-156683. This article points out that landlords may evict tenants who call the police to report domestic violence perpetrators—if they have a "zero tolerance" crime policy. In addition, eviction records can make people ineligible for public housing, making it harder to escape abuse.
7. https://ourworldindata.org/covid-deaths.
8. https://www.bls.gov/charts/employment-situation/civilian-unemployment-rate.htm and http://www.oecd.org/newsroom/unemployment-rates-oecd-update-june-2020.htm.
9. https://prospect.org/coronavirus/american-farmworkers-essential-but-unprotected/ by David Bacon. https://www.americanprogress.org/article/protecting-undocumented-workers-pandemics-front-lines-2/ points out that even U.S. citizens married to an undocumented worker were unable to access the payments provided to most adults under the CARE Act.
10. https://www.studyfinds.org/healthcare-workers-three-times-more-likely-contract-coronavirus/.
11. https://www.gao.gov/assets/gao-20-427.pdf, p. i. Also 12/9/21: https://www.wsj.com/articles/the-cost-of-rent-is-rising-fast-heres-what-americans-can-do-about-it-11639045803?mod=djem10point.
12. https://www.consumerfinance.gov/about-us/newsroom/new-report-from-consumer-financial-protection-bureau-finds-over-11-million-families-at-risk-of-losing-housing/. 19% of adult renters (13.5 million) were behind in payments and over 1/3 (37%) reported having no employment income and were worried about eviction… see more below https://evictionlab.org/eviction-tracking/ and https://unitedwaynca.org/stories/effect-pandemic-homeless-us/.
13. https://www.wsj.com/articles/covid-19s-toll-on-u-s-business-200-000-extra-closures-in-pandemics-first-year-11618580619.
14. https://www.jpmorgan.com/insights/research/cares-act.
15. https://www.bloomberg.com/news/articles/2020-05-02/the-non-bailout-how-the-fed-saved-boeing-without-paying-a-dime.
16. Lydia DePillis has several articles on the PPP program, its poor design and use of funds at: https://www.propublica.org/people/lydia-depillis. Also, the Treasury set aside $75 billion for a Main Street Lending Program, administered by the Federal Reserve. https://www.investopedia.com/main-street-lending-program-4802310. This program was to support loans to companies with fewer than 15,000 employees or less than $5 billion in revenue last year. The Fed agreed to buy 95% of any

loan that a bank made to these "small and midsize" companies as long as the borrower met certain conditions.

17. https://blog.dol.gov/2021/01/11/unemployment-benefits-answering-common-questions.
18. https://www.gao.gov/blog/struggling-homeowners-get-extension%2C-did-covid-19-housing-protections-really-work.
19. https://www.cbpp.org/research/poverty-and-inequality/american-rescue-plan-act-will-help-millions-and-bolster-the-economy#schools.
20. https://ourworldindata.org/grapher/public-trust-in-government.
21. For example, https://time.com/5851633/best-global-responses-covid-19/.
22. The most detailed account of what happened in the storm's aftermath, with the data given below and much more, can be found in the 582-page bipartisan report of the U.S. House of Representatives (2006). http://katrina.house.gov/full_katrina_report.htm. "It remains difficult to understand how government could respond so ineffectively to a disaster that was anticipated for years, and for which specific dire warnings had been issued for days. This crisis was not only predictable, it was predicted," p. xi. Number stranded (U.S. House of Representatives, Bipartisan Report 2006: 116): An estimated 70,000 persons were left in the city when the hurricane hit. http://www.gao.gov/new.items/d06903.pdf. The Coast Guard alone rescued 33,500 people. It moved many of them to higher ground where no food or water was available (Bipartisan, p. 231). 9792 MRE's and 13,440 liters of water were at the Superdome (Bipartisan, pp. 64, 118). Thurs night evacuation was requested of Superdome for next morning (p. 119). Flood waters required use of high clearance vehicles to deliver supplies (p. 120). DHS info wrong at times—as of Wed, DHS didn't know about loss of electricity at Superdome on Monday (p. 117), for example. http://www.livescience.com/11148-hurricane-katrina-history-numbers-infographic.html and http://www.truth-out.org/archive/component/k2/item/60946:bill-quigley--six-months-after-katrina-who-was-left-behind. Food never arrived at Superdome: http://ftw.usatoday.com/2015/08/refuge-of-last-resort-five-days-inside-the-superdome-for-hurricane-katrina. Fatalities: https://fivethirtyeight.com/features/we-still-dont-know-how-many-people-died-because-of-katrina/ cites Ezra Boyd (2006). "Preliminary Summary Report of Hurricane Katrina Deceased Victim Recovery Locations in Louisiana" (PDF), LSU Hurricane Public Health Center. "In August 2006, Louisiana counted 1464 victims, with 135 people still *missing*." Among federal agencies, the National Oceanic and Atmospheric Administration has been the primary one focused on determining how many people died because of Katrina regionwide. Housing damage: Of the 252,000 owner-occupied housing units in

metropolitan area NO that sustained damage during Hurricane Katrina, 90% have undergone repair. 577,000 total housing stock. https://portal.hud.gov/hudportal/HUD?src=/press/press_releases_media_advisories/2010/HUDNo.10-191. HUD data, which includes damage from Hurricanes Wilma and Rita that year, show that in the City of New Orleans proper, 71% of occupied housing suffered damage. 134,344 units in Orleans Parish. Business bankruptcies: https://www.law360.com/newyork/articles/9298/bankruptcy-concerns-hover-over-new-orleans. By Erin Marie Daly New York (August 28, 2006)—One year after Hurricane Katrina …, small businesses and taxpayers are still struggling. Approximately 81,000 businesses in Louisiana were damaged in the storm, and although 75% have since reopened, around 19,000 businesses have been permanently shuttered, according to the Census Bureau. *Facts for Features: Katrina Impact*, Allison Plyer, August 26, 2016.

23. See also note 19. https://georgewbush-whitehouse.archives.gov/reports/katrina-lessons-learned/ is also an important source from 2006. GAO. 1993. *Recent Disasters*. https://www.gao.gov/products/t-rced-93-4 Hurricane Andrew in south Florida in 1992: In that disaster, FEMA was too reactive—waiting for state and local governments to identify what they needed and then ask for it—when the hurricane had clearly overwhelmed the ability of state, local, and voluntary agencies to adequately provide victims with essential services, such as food and water, within 12–24 hours. The GAO described the [then] strategy of FEMA for responding to such disasters as deficient because it lacked provisions for the federal government to (1) immediately, comprehensively assess damages and the resulting needs of disaster victims and (2) provide food, shelter, and other essential services when victims' needs outstripped what state, local and voluntary agencies could themselves provide (p. 113). City didn't ask for food etc. for Superdome until the 'last minute' (p. 220s) lots of detail, e.g. requests made were not followed up with paperwork so were ignored, number of troops by day… (p. 320). FEMA food and supplies outside city and not easily accessible (p. 347) requests lost, mishandled etc. A search on word "stopped" yields stories about gross misbehavior of police. GAO questions: Why was coordination and information sharing between local, state, and federal governments so dismal.

    - Why situational awareness was so foggy, for so long.
    - Why all residents, especially the most helpless, were not evacuated more quickly.
    - Why supplies and equipment and support were so slow in arriving.

- Why so much taxpayer money aimed at better preparing and protecting the Gulf coast was left on the table, unspent or, in some cases, misspent.
- Why the adequacy of preparation and response seemed to vary significantly from state to state, county to county, town to town.
- Why unsubstantiated rumors and uncritically repeated press reports—at times fueled by top officials—were able to delay, disrupt, and diminish the response.
- And why government at all levels failed to react more effectively to a storm that was predicted with unprecedented timeliness and accuracy.

24. https://www.theguardian.com/us-news/2015/aug/16/hurricane-katrina-new-orleans-looting-violence-misleading-reports source for Blanco's comments on looters. Compilation of FEMA's Rejections of Qualified Help. http://www.WantToKnow.info/femafailureskatrina. Hurricane Katrina: Compilation of FEMA's Rejections of Qualified Help. Sample links: FEMA refuses hundreds of personnel, dozens of vehicles—*Chicago Tribune*, 9/2/05. http://www.chicagotribune.com/news/local/chi-050902daley,1,2011979.story. FEMA won't let Red Cross deliver food—*Pittsburgh Post-Gazette*, 9/3/05. http://www.post-gazette.com/pg/05246/565143.stmFEMA fails to utilize Navy ship with 600-bed hospital on board—*Chicago Tribune*, 9/4/05. http://www.chicagotribune.com/news/nationworld/chi-0509040369sep04,1,4144825.story. FEMA won't accept Amtrak's help in evacuations—*Financial Times*, 9/5/05. http://news.ft.com/cms/s/84aa35cc-1da8-11da-b40b-00000e. FEMA turns back Wal-Mart supply trucks—*New York Times*, 9/6/05. http://www.nytimes.com/2005/09/05/national/nationalspecial/05blame.html on Saturday FEMA (they) cut the (Jefferson) parish's emergency communications line, leading the sheriff to restore it and post armed guards to protect it from FEMA, Mr. Broussard (president Jefferson Parish) said. Article doesn't explain why water from Walmart turned away. FEMA blocks 500-boat citizen flotilla from delivering aid—*News Sentinel*, 9/8/05. http://www.fortwayne.com/mld/newssentinel/news/editorial/12595873.htm. FEMA: "First Responders Urged Not to Respond" Unless Dispatched—FEMA's own website. http://www.fema.gov/news/newsrelease.fema?id=18470. Also see notes 12–17.
25. Also https://www.nbcnews.com/storyline/puerto-rico-crisis/delayed-response-puerto-rico-has-echoes-katrina-some-n805861, https://www.nejm.org/doi/full/10.1056/NEJMsa1803972, and https://www.cbo.gov/publication/44601. The Budgetary Impact of the Federal Government's Response to Disasters, Daniel Hoople, 9/23/2013. The

CBO estimates that 93%—or $96 billion—of the total appropriations, or budget authority, for the 2005 hurricanes has been expended … more than $100 billion targeted to disaster-affected areas. The largest share of those appropriations ($50 billion) was for the FEMA. To the Department of Housing and Urban Development (HUD), $20 billion, the U.S. Army Corps of Engineers ($16 billion), and the Department of Defense ($9 billion). … Insured losses from the three storms taken together were estimated by insurance companies at $57 billion, not including $17 billion in claims paid by the National Flood Insurance Program. http://www.nola.com/katrina/index.ssf/2015/08/federal_governments_120_billio.html.

26. http://www.nola.com/environment/index.ssf/page/washing_away_2002.html.
27. http://www.nola.com/katrina/index.ssf/2015/08/federal_governments_120_billio.html?.
28. http://www.politico.com/magazine/story/2015/08/katrina-ten-years-later-michael-brown-121782.
29. See note 1.
30. http://www.espn.com/espnw/features/article/13514606/hurricane-katrina-superdome-new-orleans-national-guard. About 3000 National Guards were deployed over the course of the crisis. See also note 1.
31. See note 1.
32. https://www.aclu.org/news/aclu-report-details-horrors-suffered-orleans-parish-prisoners-wake-hurricane-katrina and https://www.vice.com/en_us/article/5gjdxn/hurricane-katrina-was-a-nightmare-for-inmates-in-new-orleans-829. One resident—a stranded publisher and professor who had taken refuge on the overpass near the prison—describes watching as prisoners were rescued from the overpass via boat, while she and other residents were left without food or water for another day or more. http://www.marygehman.com/surviving-hurricane-katrina.html.
33. http://www.nytimes.com/2009/08/30/magazine/30doctors.html?pagewanted=all, 2 a.m. on Wednesday, August 31. Memorial's backup generators sputtered and stopped on the seventh floor, critically ill patients began suffering the consequences. Alarm bells clanged as life-support monitors and ventilators switched to brief battery reserves.… Volunteers began carrying the LifeCare patients who relied on ventilators down five flights of stairs in the dark.
34. See notes 1 and 2.
35. Scott McClellan: "I think you heard from the President earlier today about his zero tolerance. We understand the need for food and water and supplies of that nature. That's why we have a massive effort underway to continue getting food and water and ice to those who are in need. There

are ways for them to get that help. Looting is not the way for them to do it." Related to Bush's visit to coast September 2. https://georgewbush-whitehouse.archives.gov/news/releases/2005/09/20050901-2.html.
36. http://www.outsidethebeltway.com/la_governor_warns_troops_will_/.
37. See note 3.
38. http://www.nbcnews.com/id/9427111/ns/us_news-katrina_the_long_road_back/t/gretna-mayor-defends-bridge-blockade/#.We5klohry70.
39. Police also added to the woes of those stranded when they closed access to a bridge out of the city at gunpoint. Gretna, the first town across the Crescent City bridge from New Orleans, initially helped refugees from the city, but their requests to FEMA for supplies and assistance went unanswered. Gretna officials in cooperation with the Department of Transportation and the Crescent Connection police closed the bridge to their city September 1.
40. http://www.espn.com/espnw/features/article/13514606/hurricane-katrina-superdome-new-orleans-national-guard.
41. http://usacac.army.mil/cac2/cgsc/carl/download/csipubs/wombwell.pdf. See also the U.S. Bipartisan Report. http://abcnews.go.com/Sports/despair-hope-national-guardswoman-reconciles-memories-katrina/story?id=33358487.
42. http://www.nola.com/crime/index.ssf/2010/04/civilian_pleads_guilty_in_danz.html. nytimes.com/2016/ and http://www.pressreader.com/usa/the-commercial-appeal/20080814/283420597419357.
43. http://www.cnn.com/2005/US/09/08/katrina.redcross/. LA DHS (Mayeaux) rebuffed Red Cross offer of food and water, saying takes time to set up to feed so many!! this was September 1, 3 days after storm!!! https://www.organicconsumers.org/news/government-denied-water-new-orleans-residents-during-katrina-relief-workers-say. A FEMA Performance Review by the White House tried to track 57 requests thru system—could only track 3 (p. 74). Spoke of food only at shelters that had been set up. See note 3 for more examples of FEMA's refusals of aid. https://georgewbush-whitehouse.archives.gov/news/releases/2005/09/20050901-2.html.
44. https://www.newsweek.com/government-response-katrina-disaster-within-disaster-118257.
45. Institute for Southern Studies Katrina report, p. 24. http://www.solvingpoverty.com/printer/ISSKatrinaYearReport-1.pdf.
46. U.S. Census Bureau, American Community Survey. See also https://www.brookings.edu/wp-content/uploads/2012/04/0114_ISSKatrina.pdf on number of persons displaced and their relocation. https://www.ncbi.nlm.nih.gov/pmc/articles/PMC4048822/ also.

47. This was not the case elsewhere in the city, except that residents of the predominately BlackLower 9th Ward were kept from their homes for three months—compared to a two-week wait in other areas. In addition, the cost of repairs...see next page.
48. Private insurance companies, surprisingly, had record profits of $45 billion in 2005 on property casualty policies, despite paying out $41 billion in claims for storms including Katrina. $18 billion went to homeowners and $21 billion to commercial enterprises. http://www.iii.org/sites/default/files/docs/pdf/HurricaneKatrinaFactFile-0320101.pdf and http://articles.latimes.com/2006/apr/05/nation/na-insure5. The industry covered virtually all of its claims and expenses with premiums earned during the year rather than with surplus funds, according to the organization's analysis. The surplus reached $427 billion. Also insured itself against losses.
49. http://www.nytimes.com/2007/09/03/us/nationalspecial/03orleans.html and https://attorneypages.com/hot/louisiana-sues-for-insurance-bad-faith.htm. The state of Louisiana estimates that homeowners have, on average, received $5700 less than insurance companies rightfully owed them (in *nytimes* article).
50. Department of Homeland Security: https://www.oig.dhs.gov/reports/audits-inspections-and-evaluations.
51. https://www.dol.gov/whd/aam/DBA_Reinstal.PDF. Davis-Bacon Act required federal contractors to pay prevailing wage.
52. http://www.law.columbia.edu/sites/default/files/microsites/career-services/Adequacy%20of%20Labor%20Law%20Enforcement%20in%20New%20Orleans.pdf (p. 3). http://www.motherjones.com/politics/2007/07/post-katrina-aftermath-absence-oversight-reconstruction-workers-became-another-casu/.
53. http://www.naomiklein.org/shock-doctrine/resources/part7/chapter20, August 28, 2015.
54. https://www.nbcnews.com/ and https://www.nbcnews.com/storyline/hurricane-katrina-anniversary/latino-workers-helped-rebuild-new-orleans-many-werent-paid-n417571. Also see Bacon (2008).
55. https://www.thenation.com/article/archive/who-killing-new-orleans/. FEMA paid Shaw $175 a square foot to install blue tarps on damaged roofs, although the tarps themselves had been provided by the government. The workers who actually hammered in the tarps were paid as little as $2 a square foot, wrote Mike Davis. "Every level of the contracting food chain, in other words, is grotesquely overfed except the bottom rung," he added. Related story: In Mississippi, a class action lawsuit forced several companies to pay hundreds of thousands of dollars in back wages to immigrant workers. Some immigrants were not paid at all. Some undocumented immigrant workers reported being

wakened in the middle of the night by their employer who told them that immigration agents were on their way; workers fled to avoid arrest.

56. Firms receiving contracts weren't necessarily experienced in doing the work for which they were hired. Klein (2008) mentions that AshBritt, the company paid half a billion dollars to remove debris, reportedly didn't own a single dump truck and farmed out the entire job to contractors.
57. K. Gotham and M. Greenberg. 2014. *Crisis Cities*. New York: Oxford University Press.
58. www.nola.com/katrina/index.ssf/2015/08/new_orleans_public_housing_dem.html.
59. American Community Survey, 2005, 106.7/436.1, 2006 47.2/212.9, 2010 92.0/338.9, 2015 88.9/379.7. https://talkpoverty.org/2016/08/26/people-poverty-work-paul-ryan-misunderstands-poverty/ and https://www.census.gov/quickfacts/fact/table/neworleanscitylouisiana/PST045216.
60. http://www.nola.com/katrina/index.ssf/2015/08/federal_governments_120_billio.html. Bruce Alpert, 8/21/15 updated 8/22/15.
61. http://usatoday30.usatoday.com/news/nation/2007-06-21-new-orleans-crime_N.htm, citing Dr. Kevin Stephens in *Disaster Medicine*. https://www.thenation.com/article/why-was-new-orleanss-charity-hospital-allowed-die/. The city set aside 37 acres for a new Louisiana State University (LSU) medical complex, razing homes and businesses. This medical center was designed to be cutting-edge and aimed at attracting higher-income clientele. What was to be a joint complex with some shared VA and LSU facilities on thirty-seven acres ballooned into two stand-alone, totally separate hospitals on sixty-seven acres.
62. www.nola.com/katrina/index.ssf/2015/08/new_orleans_public_housing_dem.html.
63. https://www.wunderground.com/hurricane/surge_wetlands.asp. About 2000 square miles of wetlands have been lost since the 1930s. Wetlands absorb storm energy and so reduce the impact of hurricanes. https://www.scientificamerican.com/article/losing-ground-southeast-louisiana-is-disappearing-quickly/.
64. http://www.nola.com/environment/index.ssf/2011/06/louisiana_is_losing_a_football.html.
65. See, for example, https://www.thenation.com/article/white-new-orleans-has-recovered-from-hurricane-katrina-black-new-orleans-has-not/ and http://understandingkatrina.ssrc.org/Gilman/.
66. http://www.slate.com/articles/news_and_politics/politics/2015/08/hurricane_katrina_10th_anniversary_how_the_black_lives_matter_movement_was.html and https://talkpoverty.org/2016/08/26/people-poverty-work-paul-ryan-misunderstands-poverty/.

67. http://www.naomiklein.org/shock-doctrine/resources/disaster-capitalism-in-action/tags/hurricane%2Bkatrina. *The Shock Doctrine*. New York: Metropolitan Books, 2008.
68. E.g. Klein (2018). The *New England Journal of Medicine* finds the number of deaths in Puerto Rico after Hurricane Maria was vastly underreported: https://www.nejm.org/doi/full/10.1056/NEJMsa1803972.
69. U.N. (2011: 51). The economic crisis increased the number of food insecure persons by about 9% in 2009, over and above the pre-crisis projected baseline increase of 2% for 2009.
70. Relevant data: Eviction Rates, Selected Cities: https://evictionlab.org/national-estimates/. Rise in Federal Deficit: https://www.thebalance.com/us-deficit-by-year-3306306#deficit-trends. Rise in Debt, Falling Revenue of State and Local Government: https://apps.bea.gov/iTable/iTable.cfm?reqid=19&step=2#reqid=19&step=2&isuri=1&1921=survey, https://www.brookings.edu/articles/state-and-local-budgets-and-the-great-recession/, and https://www.census.gov/library/publications/2011/compendia/statab/131ed/state-local-govt-finances-employment.html. Summary shows negative savings.
71. See https://www.forbes.com/sites/melaniehaiken/2014/06/12/more-than-10000-suicides-tied-to-economic-crisis-study-says/#3ee6bd217ae2, https://www.bloomberg.com/news/articles/2016-04-05/how-the-great-recession-increased-domestic-abuse, and https://www.ncbi.nlm.nih.gov/pmc/articles/PMC3705414/.
72. https://data.bls.gov/timeseries/LNS14000000 for monthly unemployment rate, seasonally adjusted https://www.bls.gov/lau/stalt_archived.htm on alternative measures of unemployment https://www.bls.gov/opub/ted/2009/ted_20090828.htm on unemployment among those 16–24. https://data.bls.gov/pdq/SurveyOutputServlet. Black unemployment: https://fred.stlouisfed.org/series/LNS14000006. 16–19 years of age: https://fred.stlouisfed.org/series/LNU04000012. 65 years and older: https://fred.stlouisfed.org/series/LNU04000097. 55 and over: https://fred.stlouisfed.org/series/LNS14024230. Unemployed 27 weeks and longer: https://fred.stlouisfed.org/series/UEMP27OV. Underemployment Rate (U-6): M. https://fred.stlouisfed.org/series/U6RATEM.
73. Americans for Financial Reform (2015: 4, 13). See also http://www.epi.org/publication/understanding_the_jobs_crisis/ which reports results of a September 2009 survey by Hart Research, which found that 44% of households had experienced job loss or cuts in pay or hours.
74. The median income of a group is the income such that half of the group has an income above the median and half has an income below the median. Americans for Financial Reform (2015: 13). Federal Reserve

Survey of Consumer Finances, Table 4, 2013 and 2016. https://www.federalreserve.gov/publications/files/scf17.pdf.
75. Federal Reserve (2017: 5, 8).
76. Wolff (2021: 48). Net worth here includes the value of housing minus mortgage outstanding.
77. Federal Reserve. 2020. *Survey of Consumer Finances*. https://www.federalreserve.gov/econres/scf/dataviz/scf/chart/#series:Net_Worth;demographic:all;population:1;units:median, net worth and income Wolff (2021: 50).
78. Americans for Financial Reform (2015: 17–18). Foreclosure data from: GAO. 2013. *Financial Regulatory Reform*. 1/13. GAO-13-180, p. 25. Foreclosure rate rose from 1% to 4+% in 2010. http://www.econ.yale.edu/~shiller/data.htm. Shiller's data set shows stock market fluctuations. http://www.tanocapital.com/news/2008/11/14/tano-capital-reseach-the-crash-of-2008-vs-1929-similarities.html gives data on the crash in prices.
79. https://www.newyorker.com/magazine/2016/02/08/forced-out. Data in GAO report: https://www.gao.gov/assets/gao-20-427.pdf, p. i. See also Matthew Desmond. 2016. *Evicted*, 344, 360. New York: Crown Publishers. https://www.redfin.com/blog/2016/12/millions-of-renters-face-eviction-why-todays-housing-market-is-partially-to-blame.html gives data on the 30 largest cities for which eviction data was available for 2014; it found that Milwaukee's rate was lower than most.
80. Several articles stated that banks had the practice of accepting 'stated income'—not checking whether or not income claims were accurate. This and other risky practices are detailed in: http://www.businessinsider.com/how-widespread-mortgage-fraud-toppled-the-housing-bubble-2010-5.
81. 'I worked in the subprime units at Countrywide and Wells Fargo during the bubble years (2004–2007). I saw this phenomenon first hand. To pretend that there is anything mysterious or unknown about the methodology of these loans is simply to add a lie on top of a lie on top of a lie,' said Michael David White April 22, 2010. http://www.nytimes.com/2010/02/28/business/economy/28gret.html?pagewanted=1&partner=rss&emc=rss and https://www.project-syndicate.org/commentary/securing-synthetic-securities. George Soros opined that CDSs serve no productive function, merely creating profit from bubbles. http://people.stern.nyu.edu/msubrahm/papers/ARFE.pdf provides academic overview of CDSs.
82. https://www.occ.gov/topics/capital-markets/financial-markets/derivatives/derivatives-quarterly-report.html.

83. http://www.marketwatch.com/story/lehman-folds-with-record-613-billion-debt?siteid=rss. In March, the New York Fed arranged a deal whereby JP Morgan acquired the bankrupt investment firm Bear Stearns, with $30 billion of the loan to buy the firm guaranteed by the Federal Reserve.
84. https://www.federalreservehistory.org/essays/fed_credit_programs. See also https://fred.stlouisfed.org/series/INTDSRUSM193N.
85. https://www.treasury.gov/initiatives/financial-stability/reports/Documents/TARPfirst-105report.pdf. The Congressional Budget Office (2009) estimated that the first $247 billion of securities purchased represented 26% ($64 billion) subsidy to the banks receiving funds. http://ritholtz.com/2009/01/cbo-report-on-tarp/. Note: The Federal Reserve, whose board of governors is appointed by the president, is able to make its decisions independently, subject to Congressional oversight. [There are staggered 14-year terms for the Board, and the majority of directors of the district Reserve banks are chosen by private banks.] https://www.theatlantic.com/business/archive/2014/02/how-the-fed-let-the-world-blow-up-in-2008/284054/. The Fed kept rates at 2.25% as of September 16. http://www.motherjones.com/politics/2009/12/real-size-bailout-treasury-fed/: lists all large programs and their funding authorizations.
86. U.S. Treasury (2021) on TARP loans.
87. https://www.osc.state.ny.us/osdc/wallst_bonuses/2010/022310.htm, $20.3 billion. http://www.osc.state.ny.us/press/releases/mar16/nyc_security_avg_bns.pdf and http://www.washingtonpost.com/wp-dyn/content/article/2009/07/22/AR2009072203687.html.
88. www.nytimes.com/2012/04/01/business/how-a-financial-products-agency-could-protect-investors.html by Gretchen Morgenson.
89. https://www.nytimes.com/2014/05/04/magazine/only-one-top-banker-jail-financial-crisis.html.
90. See Table 4 of Office of the Comptroller of the Currency annual reports on derivatives at https://www.occ.gov/topics/capital-markets/financial-markets/derivatives/derivatives-quarterly-report.html.
91. www.nytimes.com/2012/04/01/business/how-a-financial-products-agency-could-protect-investors.html. Posner and Weyl discuss by Gretchen Morgenson. https://psc-cuny.org/clarion/march-2012/how-business-elites-looted-private-sector-pensions details why pensions also need protection.
92. https://www.brookings.edu/blog/up-front/2017/06/05/a-primer-on-dodd-franks-orderly-liquidation-authority/. Liquidation Authority Orderly fund not funded until *after* failure.
93. "The banking system, for example, has been strengthened with tougher capital and liquidity requirements. But the underlying cause of the recent

financial crisis, like many before it, remains a threat: the ever-present potential for a real estate bubble. "The financial crisis in retrospect", *Forbes*, 9-19-2013. http://www.nytimes.com/2011/10/14/business/new-dodd-frank-rules-muddled-by-congress-that-wants-it-both-ways.html?_r=1&src=recg exemptions to the Volcker rule against proprietary trading.

94. http://www.pewtrusts.org/~/media/assets/2015/01/fsm_balance_sheet_report.pdf. On wages: Council of Economic Advisers to the President (2017: Table B15). The median home price only returned to its 2007 level in 2013. https://www.census.gov/construction/nrs/pdf/uspriceann.pdf. See also https://www.census.gov/content/dam/Census/programs-surveys/ahs/publications/Drowning_in_Debt.pdf.
95. See https://www.insiderintelligence.com/insights/largest-banks-us-list/ for 2022 data.
96. U.S. Bureau of Economic Analysis, National Income Accounts, Tables 6.16C, 6.16D and 6.17D. https://apps.bea.gov/iTable/index_nipa.cfm.
97. U.S. Bureau of Economic Analysis, National Income Accounts. Tables 6.5. https://apps.bea.gov/iTable/index_nipa.cfm and https://www.theatlantic.com/business/archive/2013/03/how-wall-street-devoured-corporate-america/273732/.
98. The ten largest U.S. banks held $11.8 trillion in assets, of the $16.8 trillion in the banking system in 2017. In comparison, the total gross domestic product generated in 2017 was about $19.4 trillion. See https://www.insiderintelligence.com/insights/largest-banks-us-list/ for 2022 data.
99. Americans for Financial Reform (2015: 13–19). Wolff (2021), appendix tables.
100. https://www.imf.org/external/np/g20/pdf/062710b.pdf, p. 9. https://www.nytimes.com/2014/05/04/magazine/only-one-top-banker-jail-financial-crisis.html.
101. See Foucault (1989), Fraser (1989), Haugaard (1997), Lukes (1974), Rorty (1992), Wartenberg (1992). Also https://literariness.org/2018/08/14/the-political-theory-of-ernesto-laclau-and-chantal-mouffe/. Nasrullah Mambrol offers a far-ranging, insightful discussion of political theory, touching on Foucault, Althusser, Derrida, Lukacs, Heidegger, Laclau, Mouffe and more.
102. Their initiatives were in line with their duties as understood by citizens and leaders followed recognized channels and processes. They used the communication channels available to them effectively, explaining the need to enter the war and the directives issued. The structures of our institutions, which concentrate power at the top, enabled a strong and unified war effort. National leaders in both government and business

were in agreement, which led to a unity of purpose. A major media blitz was built on the shared belief, after Pearl Harbor, that the nation faced an imminent threat to its security. With the public behind the war effort, any opposition to war plans was weak.

103. "Nature never intended the fair sex to become players of wind instruments… women cannot possibly play brass instruments and look pretty, and why should they spoil their good looks?" Gustave Kerker. "Opinions of Some New York Leaders on Women as Orchestral Players," *Musical Standard 21* (April 2, 1904). It wasn't until the 1970s that blind auditions in the first rounds of tryouts dramatically increased the numbers of women in orchestras. Goldin, Claudia and Cecilia Rouse. 2000. "Orchestrating Impartiality: The Impact of" Blind "Auditions on Female Musicians." *The American Economic Review* 90 (4): 715–741.
104. Many that believe, for example, that all citizens should be able to vote online, securely, on any major changes proposed in federal taxes or on the allocation of expenditures in the federal budget. There are many obstacles, largely political, blocking this vision from serious consideration.

References

Americans for Financial Reform. Updated 2015. *Cost of the Crisis*. https://ourfinancialsecurity.org/policy-materials/afr-briefing-paper-cost-of-the-crisis-taking-stock-on-the-fourth-anniversary-of-the-financial-meltdown/.

Bacon, D. 2008. *Illegal People: How Globalization Creates Migration and Criminalizes Immigrants*. Boston: Beacon Press. Chapter 6 on Mississippi and Katrina.

Congressional Budget Office. 2009. *The Troubled Asset Relief Program: Report on Transactions Through December 31, 2008 (Report)*, January.

Council of Economic Advisers to the President. 2017. *Economic Report of the President*. Washington, DC: Government Printing Office.

Federal Reserve. 2017. *2016 Survey of Consumer Finances Chartbook*. https://www.federalreserve.gov/econres/files/BulletinCharts.pdf.

———. 2020. *2019 Survey of Consumer Finances*. Interactive Charts. https://www.federalreserve.gov/econres/scf/dataviz/scf/chart/#series:Net_Worth;demographic:inccat;population:all;units:median.

Federal Reserve at St. Louis.

Foucault, Michel. 1989. *The Archaeology of Knowledge*. London: Routledge.

Fraser, Nancy. 1989. *Unruly Practices*. Minneapolis: University of Minnesota Press.

Goldin, Claudia, and Cecilia Rouse. 2000. Orchestrating Impartiality: The Impact of "Blind" Auditions on Female Musicians. *American Economic Review* 90 (4): 715–741.

Gotham, Kevin, and Miriam Greenberg. 2014. *Crisis Cities*. New York: Oxford University Press.

Government Accountability Office (GAO). 2013. *Financial Regulatory Reform*. GAO 13-180. Washington, DC: Government Printing Office. https://www.gao.gov/products/gao-13-180.

Haugaard, Mark. 1997. *The Constitution of Power*. New York: St. Martin's Press.

Hurd, Michael, and Susann Rohwedder. 2010. Effects of the Financial Crisis and Great Recession on American Households. Working Paper 16407, September. National Bureau of Economic Research. https://www.nber.org/papers/w16407.

International Monetary Fund. 2010. *A Fair and Substantial Contribution by the Financial Sector*. Final Report to the G-20. https://www.imf.org/external/np/g20/pdf/062710b.pdf.

Journal of Economic Literature. 2012. Forum on What Economists Should Read About the Financial Crisis. *Journal of Economic Literature* 50 (1): 128–178.

Kerker, Gustav. 1904. Opinions of Some New York Leaders on Women as Orchestral Players. *Musical Standard 21*, April 2.

Klein, Naomi. 2008. *Shock Doctrine*. New York: Metropolitan Books (Henry Holt). http://www.naomiklein.org/shock-doctrine/resources/part7/chapter20.

———. 2015. *Change Everything or Face a Global Katrina*, August 28. https://naomiklein.org/change-everything-or-face-global-katrina/.

———. 2018. *The Battle for Paradise*. Chicago: Haymarket Books.

Lukes, Steven. 1974. *Power: A Radical View*. London: Macmillan.

Peterson, V. Spike. 2002. Rewriting (Global) Political Economy. *International Feminist Journal of Politics* 4 (1): 1–30.

Rorty, Amelie O. 1992. Power and Powers. In *Rethinking Power*, ed. Thomas Wartenberg, 1–13. Albany: State University of New York Press.

Saez, Emmanuel, and Gabriel Zucman. 2020. Trends in U.S. Income and Wealth Inequality. National Bureau of Economic Research Working Paper 27921, October. https://www.nber.org/papers/w27921.

Scientific American. 2022. *Special Issue: How COVID Changed the World*, March, vol. 326, no. 3. NY.

Unemployment Rates. https://fred.stlouisfed.org/series/UNRATE.

Unemployment Rates by Category. https://fred.stlouisfed.org/categories/32447.

United Nations. 2011. *The Global Social Crisis*. New York: United Nations. http://www.un.org/esa/socdev/rwss/docs/2011/rwss2011.pdf.

U.S. Department of Agriculture, Food and Nutrition Service. Number of Persons on SNAP. https://fns-prod.azureedge.net/sites/default/files/resource-files/SNAPsummary-8.pdf.

U.S. House of Representatives. 2006. *A Failure of Initiative: Final Report of the Select Bipartisan Committee to Investigate the Preparation for and Response to Hurricane Katrina*, February 15. Washington, DC: Government Printing Office. http://katrina.house.gov/full_katrina_report.htm.

U.S. Treasury. 2021. *Troubled Asset Relief Program.*

Wartenberg, Thomas, ed. 1992. *Rethinking Power*. Albany: State University of New York Press.

Wolff, Edward. 2021. Household Wealth Trends in the United States, 1962 to 2019. Working Paper 28383. Cambridge, MA: National Bureau of Economic Research. http://www.nber.org/papers/w28383.

CHAPTER 3

Power: A Class Act

A nation's laws and practices shape the course of its destiny.

As we have seen, Hurricane Katrina, the financial crisis of 2007–2009 and the coronavirus pandemic revealed not merely cracks but deep fissures in our polity. Events have repeatedly shaken our faith in the nation's leadership. This is reflected in the drop in polling numbers for most major institutions, from big business to the presidency and the Supreme Court. Confidence in the U.S. president pre-Katrina, 2003, was at 78%, but by 2006, the year after Katrina, had dropped to 50%, and since 2017 has been under 50%, according to Pew Research.[1] Another poll by Pew found a similar drop in the trust in the federal government: in 2020 only 20% trusted the federal government to "do the right thing," down from 77% in 1964. As to other institutions, only 11% of those polled by Gallup expressed "quite a lot of confidence" in Congress, a lower percentage than in big business (23%, down from 34% in 1975).[2]

We are worried. We wonder whether our leaders will be capable of doing what is needed during the crises that lie ahead. The global pandemic is only the latest crisis to meet with a dismal response. In addition, global warming, increasing income inequality, environmental degradation, tensions among nations with nuclear capabilities, and the rise of hate groups around the world present imminent threats to world

M. T. Hill, *Building a Solidarity Society*,
https://doi.org/10.1007/978-3-031-07349-6_3

stability. Yet national initiatives to take on these issues proactively have been sadly inadequate; the price we will pay will undoubtedly be high.

The moral challenges we face are the equals of those of earlier times. But the consequences of failure are more dire; failure would leave in its wake a devastated planet with diminished prospects for coming generations of humanity. Only a transformation of our societies can provide the foundation for a community of humanity dedicated to advancing the collective good. This shift in direction is vital; it demands an end to the practices that protect and privilege the powerful few who are jeopardizing our common future.

How do we do this? The concentration of power in today's ruling class is at the crux of the problem. A ruling class by its very nature strives to maintain its position, typically even at the expense of the collective good. In the U.S. and most countries, power flows from the ownership of, and control over, wealth. Under capitalism, upper-class status is based primarily on the wealth created through corporates. This social reality is at odds with a core value of the country: the belief in democracy and in rule by the people. The deep belief in democracy shared by Americans offers hope that we can build the collective will to end the domination of the country's wealth and political system by an elite few.

No ruling elite is invincible; power yields to greater power; and this greater power can be built by a coalition of those with an alternative, more humane vision for society. Building this coalition is the central task of our time; this coalition must be strong enough to prevail in the inevitable confrontation with those protecting "the way things are done," that is, with those benefiting from the way power is structured.

There is much we know about how to create a shared commitment that can mobilize people for difficult tasks. Related to this, there is much we know about the wealthy, about the institutional framework that supports them, and about the interlocking factors that enable their continued dominance. Rulers have fallen from power in the past and history tells us much about why. History in fact provides highly relevant insights into how political transformations leading to greater democracy have occurred. So, we look now at the power of the wealthy in the U.S. and then, in this and the next chapter, at paths that can lead to the creation of a countervailing power capable of creating a more egalitarian, sustainable, and democratic society.

Politics Today

Voters increasingly recognize the devastating impact of the policies and practices now in place. The COVID-19 pandemic has made clear the dire consequences of a weak and unraveling social safety net. Even before the pandemic hit, inadequate health care, low wages, institutionalized discrimination, and an ever-more unequal distribution of wealth plagued the country. About half of families in the U.S. are just one crisis away from poverty. Many young adults with college degrees are crushed by student debt for decades after graduation. Others without degrees enter the workforce in low-paying jobs and must work years before landing a job that pays a living wage, typically a wage that is lower than that enjoyed by their parents. Thousands of workers have lost expected retirement pensions as firms are sold and obligations to employees voided. As things stand, our economic, political, judicial, educational, health care, and other systems fail us too often. The achievements of other countries in health care, childcare, and worker protections make it clear we can do better.

Presidential candidates, of course, are quick to recognize the widespread discontent with the current state of affairs. Trump, in his runs for the presidency, pitched his candidacy to voters through extravagant promises to solve economic and social problems. "I will make America great again," was his 2016 theme. He even promised health care for all: "I am going to take care of everybody, I don't care if it costs me votes or not. Everybody's going to be taken care of, much better than they're taken care of now … The government's going to pay for it."[3] His promises changed in 2020, after the coronavirus pandemic and the Black Lives Matter demonstrations, to law and order, a vaccine for COVID-19 and of course a prosperous economy.[4]

Obama before him was measured in his promises, but held up a grand vision.[5] "Yes we can" was his slogan and he applied it to a wide range of policy goals. The Democratic field of Presidential hopefuls in 2020 raised the expected bread-and-butter issues but also climate change. In addition, Bernie Sanders and Elizabeth Warren took on income inequalities and corporate power. Joe Biden, who emerged as the Democratic nominee, pledged to build on the unrealized promise of the Obama legacy and after the election, under pressure from progressives, unexpectedly shifted to the left from his initial middle-of-the-road positions.[6]

What Do the Wealthy Want?

But, after decades of political promises of good jobs, affordable health care, infrastructure repair, and more, most voters no longer think that the election of an individual to the presidency will, in and of itself, suffice to solve the problems we face. After all, the persons heading our institutions from the presidency on down have changed many times over the years, yet results, under very different leaders, have not changed much for mainstream Americans. Nor has environmental destruction stopped.

This is not surprising, given that our institutions are structured to empower the wealthy few, who are largely able to set and pursue the goals of their choice. And they choose to protect their interests as they set policies that should be serving the best interests of all. To see specifically how conflicts of interest affect the course of the country, we answer the following questions: How wealthy are the wealthy? What goals do they pursue that differ from those of the rest of us? What do they in fact do that impacts us all? We then look at the widely shared beliefs about the prerogatives attached to wealth that must be revised if we are to claim our voice in government and make our democracy truly representative of the public interest.

How wealthy are the wealthy? The wealthiest 1% of U.S. households had an average net worth of $27.6 million in 2019.[7] They held 45% of all financial assets, excluding the value of primary residences, while the top 10% owned over 90% of all financial securities, stocks, and mutual funds.[8] Those in the Forbes 400 in 2020 each had a net worth of over $2.1 billion![9] In contrast, the net worth of America's statistical middle—the nation's middle 20% of income—earning households—has dropped from a median of $109,120 in 2007 to $95,620 in 2019, a 12% fall.[10]

Many in Middle America are one crisis away from losing their life savings. About 80% of American workers say they're living paycheck to paycheck, according to a 2017 report by the employment website CareerBuilder.[11] Women are particularly vulnerable: 81% report living paycheck to paycheck, compared with 75% of men. A 2018 Federal Reserve report found that "40% of U.S. families say they cannot cover a $400 emergency expense, or would do so by borrowing or selling something."[12]

Would $200,000 Help Your Family?
Conventional wisdom has it that redistributing wealth from the top 1% to the bottom 90% wouldn't make much difference, but that is far from true. There were about 1.3 million households in the top 1% in 2019, with an average net worth of $27.6 million, for a total net worth of $35.9 trillion.[13] With a U.S. population of 328 million, every man, woman, and child in the country would receive $109,451 if this wealth were divided evenly! Twice that, if the money went only to the bottom 50%.

If we look at income rather than wealth, those in the top 1% of households had an average income of $1.96 million in 2018.[14] The top 0.01% and 0.001% had average annual incomes that were much higher: $35 million and $152 million, respectively, in 2015.[15] If the income of the top 1% were distributed equally among the bottom 50% of households, each would receive over $39,500 annually.

What about our Congressional representatives? Most members of Congress have a net worth of over $1 million, with Senators wealthier than Representatives on average. Incumbents do especially well. A study of 616 incumbents calculated that "The average increase in net worth in the top 100 was 114% per year."[16] In terms of income, as opposed to wealth, the numbers confirm that members of Congress still rank high, given that the lowest salary of Senators and Representatives in Congress is $174,000 (2020).[17]

Even members of Congress who are not millionaires are friends with, and rely heavily on, the 1%. They must, after all, finance election campaigns that cost in millions of dollars. And the political parties to which they belong also rely on the support of the wealthy. The dependence of politics on big money cements the influence of top wealth-holders on the political process. And, in addition, the wealthy are more generous in their financial backing of politicians who largely share their views on policy matters, which gives those politicians an edge in campaigns.

What do the wealthy want that differs from the mainstream? There are many sources of wealth and a wide range of concerns among the wealthy. No matter the source of wealth, however, simply being wealthy affects a person's economic security, social status, friendships, and choices. Being wealthy influences personal agendas and governmental policy preferences, especially those related to finance and taxes. In addition, the wealthy have higher levels of educational attainment, more extensive travel experience, and friends and neighbors who are also wealthy. These and other factors impact their opinions.

The following paragraphs present some rather non-controversial generalizations about the impact of wealth on a person's priorities. Keep in mind there are always individual exceptions to the rule.

Economic security. Like most of us, the wealthy act to protect or improve their social and economic status. They value their ties to business, whether due to the high-paying jobs they hold, their ownership of successful firms, or their financial investments. They often belong to several business associations and may serve on the boards of different corporations. Interlocking directorates are common: the business elite often serve on the board of more than one business or nonprofit.

The wealthy value the market economy, which is the source of their wealth. Even business owners who willingly accept their broad social and environmental responsibilities strongly support the laws and market mechanisms that protect their interests. They are likely to resist change that would reduce their control over decisions made within their businesses: for example, they would likely oppose the inclusion of stakeholders from the community or employee representatives in their decision-making discussions.

They understand the concerns of their peers and speak their language. They respect success, and believe it brings responsibilities and privileges. They believe that profits are the life-blood of business and recognize that the "big boys" play a cut-throat game in which higher profits may rest on the employment of part-time workers, low-cost contract labor, and minimal compliance with federal regulations.

Policy preferences. The wealthy are in personally secure economic positions. The current system has worked for them, and for their friends and neighbors. They are not suffering from a lack of health care, unemployment, high levels of student debt, poor pensions, or inadequate housing. These problems do not afflict them. Their policy preoccupations instead lie first with the goals of the businesses they run, those of the associations to which they belong, and with the obstacles facing their relatives and colleagues. They also want a strong and prosperous country. They are aware of the power they can exercise over politicians and policymaking, and they use it. They use it to advance tax and expenditure policies that address their concerns. Those in the top 1% are especially active.

Political Activity of the Wealthiest Americans
A Russell Sage Foundation study (Page et al. 2013) examined the political activity of the wealthiest 1% in the Chicago area. The average income of the 83 respondents was over $1 million. Most attended to politics "most of the time" (84%), most contributed money (68%), and almost all voted in the last presidential election (99%). Smaller percentages attended political meetings, speeches or dinners (41%), or helped solicit or bundle contributions (21%). Forty percent personally contacted the office of their congressperson on issues they cared about. They discussed politics most days. Most identified as Republicans, 58%, and 27% as Democrats. They were less concerned about social issues than about economic issues affecting them personally.[18]

Status, social norms, and networking. The wealthy are linked by more than common interests. Social and business networks build bonds that enable them to interact directly and to strategize for goals they share as a class. The wealthiest of the wealthy get together and relax at a few very select venues across the country (Domhoff [2020] names several). Just a bit farther down the social ladder, the affluent mingle at yacht clubs, golf clubs, and similar gathering spots. In addition, they are concentrated in exclusive neighborhoods. Their children attend select schools and colleges where they learn appropriate behaviors among their peers regarding leisure activities, travel, cars, clothing, and political attitudes. In short, there are many ways to learn the opinions of peers and to develop networks.

Political preferences. Commonalities linking the wealthy result in priorities that differ from those of mainstream America. For the most part, the wealthy are financially and economically conservative. They support private enterprise, minimal government regulation, and prefer that federal budgets focus on such public goods as national security and infrastructure rather than on social goods. The wealthy may be more conservative in terms of economic policy than is the norm, but, perhaps linked to their higher levels of education, they are also more tolerant of gays, immigrants, and high profits than is the typical middle-class conservative.[19] And they are more likely to favor keeping abortion legal (Gilens 2012: 110).

Although studies of the preferences of the wealthy do not use the same definition of "wealthy" and do not ask the same questions, most confirm that the public tends to be more strongly in favor of policies that improve the economic security of the middle class than are the affluent. For example, in one study of persons with incomes of over $1 million, only 40% of respondents thought the minimum wage should be high enough to keep a full-time worker and family above the official poverty line, while 78% of the public favored such a policy (Callahan and Cha 2013: 5). Just as striking, only 19% of them believed that the government should help ensure that everyone who wanted to work could find a job, compared to 68% of the general public. Respondents to another survey of the top 10% of households (incomes over $120,000 in 2019 dollars) thought that the capital gains tax (already lower than the income tax at 15% for incomes under $400,000) should be cut further and, unlike the general public, they also preferred that the government shift from the social security system to private stock accounts. They were also less concerned about rising income inequality.[20]

These policy preferences are, without doubt, linked to the comfortable financial situation of the well-to-do. Their preferences, however, do not translate into easy predictions about any specific law under consideration by Congress or state legislatures. While the wealthy tend to be more conservative and more Republican than the typical voter, they will not necessarily support any given piece of conservative legislation. They are more likely to get involved with bills addressing economic issues rather than social policy.[21]

Another relevant point: those in the top 0.1% are generally more conservative than those in the top 10% of households, and their impact can be greater. Even a single billionaire can prevent progress on a popular piece of legislation if he or she is able to influence one key Congressperson such as the chair of the relevant committee. See the discussion below of the fate of a proposal before the federal Securities and Exchange Commission.

Some findings about the affluent are surprising. A survey by Gilens (2012: 110–114) found that the wealthiest 10% favored higher taxes on top income earners! And that they favored legal protections for gays and allowing gays in the military. Based on 2016 data, more persons with incomes over $150,000 were members of the Democratic than of the Republican party, although the percentage of Democratic was lower than that of the voting public overall.

A Plutocrat Speaks on Income Inequality
"Do we feel bad for the growing gap between the rich and the poor in the U.S.? Of course not, we celebrate it...God bless income disparity and those who have succeeded.....Income disparity? Feh! What we despise is government that imposes rules that prohibit or make it difficult to make even more money; to employ even more people; to give even more sums to the charities of our choice." Plutocrat Dennis Gartman in Freeland (2013: 249).

Information on the political position of the top 10% or the top 1% on every piece of legislation proposed is not available. Polls only address a few of the bills being considered by Congress at a given time, and the response rate can be problematic. In addition, few polls ask respondents to report their income bracket. But it is not difficult to determine the political stand of the business associations and other organizations supported by the wealthy. The positions of these organizations on legislation impacting their members are readily available at their websites, and information about their registered lobbyists is also accessible.

How the wealthy impact government. Tracking the effectiveness of lobbyists for big business, then, is one way to begin tracking the influence of the wealthy on laws and regulations affecting business, consumers, and workers.

Lobbying pays off; public loses. The record shows that discrepancies between what the public wants and what business favors are common. For example, the public favors a higher minimum wage, free college tuition, health care for all, and paid family leave—goals that remain far from becoming reality. On the other hand, business lobbyists have a relatively high rate of success. A 2014 study by the political scientists Martin Gilens of Princeton and Benjamin I. Page of Northwestern tracked how well the preferences of various groups (average Jo's, AARP, the Insurance Agents Association, the Chamber of Commerce, etc.) predicted the way that Congress and the executive branch would act on 1779 policy issues over a span of two decades. The results were shocking. Gilens and Page found that business succeeded in getting their favored policies adopted about half of the time, and in stopping legislation to which they were opposed nearly all of the time. But, they also conclude: "the preferences of the average American appear to have only a minuscule, near-zero, statistically non-significant impact upon public policy" (Gilens and Page 2014: 575). And, they add, the same lack of effect is true for mass-based interest groups like the AARP (Gilens and Page 2014: 565). They also find that, whether Congress is dominated by Democrats or by Republicans, legislators are more responsive to the top 10% than to other income groups (study covers 1964–2006). These results support theories claiming Economic-Elite Domination of U.S. politics.

Policies that benefit business. What are some of the policies that have benefited big business and the wealthy? Studies of welfare for the rich detail many of the advantages ceded to industries and high-income individuals. Industry-specific advantages include tax breaks—oil, defense, and manufacturing research. Many provisions of the individual income tax benefit higher income groups disproportionately, including the lower tax rate on capital gains versus wage income. Other benefits include low-cost access to natural resources, patent protections, low-interest federal loans, trade missions promoting U.S. businesses, preferential treatment during times of crisis, including through bailouts, and so on. In many cases, regulations passed are closer to those preferred by business than to those proposed by consumers and workers e.g. truth in advertising rules, bankruptcy laws, requirements for filing lawsuits, consumer and worker protections, environmental regulations, overtime regulations, union organizing regulations, whistleblower protections, and more.[22]

Large financial institutions are among the corporations with hefty political clout and they use it with great success. Financial regulations have far-reaching effects, given the central role that money plays in the economy. After the banking crisis, Congress faced vociferous opposition to proposed restrictions on big banks—from financiers and their clients who had made millions through risky, yet ultimately federally protected, investments. Although the public wanted to ensure that there would not be another similar crisis, the legislation that finally made it through Congress was so watered down that today the risk exposure of big banks, as reported by the Treasury, is as great as it was in 2008 just before the financial crisis.[23] The political weight of financiers has long been part of the American political scene.

"A highly complex set of laws and exemptions from laws and taxes has been put in place by those in the uppermost reaches of the U.S. financial system," explains a former investment manager. It allows them to protect and increase their wealth and significantly affect the U.S. political and legislative processes. "Ordinary citizens in the bottom 99.9% are largely not aware of these systems, do not understand how they work, are unlikely to participate in them, and have little likelihood of entering the top 0.5%, much less the top 0.1%. Moreover, those at the very top have no incentive whatsoever for revealing or changing the rules," this manager concludes.[24]

The Securities and Exchange Commission (SEC), which oversees both financial and nonfinancial corporations, is an agency that often faces intense political pressure from finance. The SEC has no obligation to respond to the expressed preference of the public but is responsive to concerns voiced by members of Congress. For example, in 2013 under pressure from the House Financial Services Committee Republicans, the SEC chair removed from consideration a regulation that would have required corporations to report political spending to shareholders—this, despite the overwhelming support for the measure reflected in the almost 700,000 comments received by the SEC (a record number). Courts can be more conservative than the SEC regarding corporate regulation: for example, a court struck down an SEC rule facilitating shareholders' ability to have their proposed nominees for a corporation's board of directors included in proxy materials.

In case there is any doubt, the clout that business has in Congress is further confirmed by the poor track record of legislation aimed at ending corporate misbehavior. The nonprofit Corp Watch follows Congressional bills directed at curbing corporate abuse. It notes that, among such bills in 2012, was one that would have denied government contracts to firms engaging in human trafficking and forced labor. It was defeated. The group also notes other reform bills that shared the same fate, including one requiring transparency in political spending, one ending subsidies to big oil, one improving arbitration fairness, and others designed to ameliorate the effects of the Citizens United decision (which recognized corporations as persons and has led to a skyrocketing of spending on political campaigns). Bills to eliminate illegal child labor practices among immigrant farmworkers in the U.S. through improving enforcement have also failed to pass. The story of Nestlé and child slavery on cocoa plantations, along with a recent court filing, is found in the accompanying sidebar.

Child Slavery on Plantations Abroad, Child Labor in U.S.
Trafficked children provide a large percentage of the workforce on cocoa plantations. They are often brutally treated. Nestlé, and other chocolate producers, back in 2001 reached an agreement with the International Labor Organization to eliminate the use of child slave labor on the plantations from which they purchase cocoa. But the problem continues, with children from several countries still working on the company's cocoa plantations.

In a statement on its website, Nestlé proclaims, "In February 2012, Nestlé became the first company in the food industry to partner with the Fair Labor Association (FLA)... to improve working conditions in supply chains. We invited the FLA to examine our cocoa supply...(its June 2012) report found that child labor is still a reality (on cocoa farms in Côte d'Ivoire)." In this same statement, rather than announcing strong steps to end this practice or even to monitor suppliers to ensure the children are well-treated and attend school, Nestlé proposed to stop child slavery on its plantations "by raising awareness and training people to identify children at risk and to intervene where there is a problem."[25]

Note: It is surprising that, despite the use of child slave labor by Nestlé's suppliers, KPMG, an international audit, tax, and advisory firm, ranked Nestlé no. 1 in corporate responsibility among food producers and among the top ten in all categories in 2013!!

A related issue is that of child labor in the U.S., which is permitted on farms with parental consent. As much as one-quarter of agricultural produce here is picked by children, some as young as six, many of whom are paid a pittance and miss many hours of school.[26] About half of these children do not graduate from high school. Poor migrant families rely on this income. Human Rights Watch notes that piece rate pay for these children can be as low as $3.60 per hour.[27] Children often work 10 hours per day on weekends or in the summer; they also work before and after school. This work can be hazardous, involving pesticides and the operation of farm machinery. Legislation to address this issue is periodically introduced in Congress.

A Child Trafficking Case (Almost) Makes It Through U.S. Courts
The Washington-based International Labor Rights Fund (ILRF) and San Francisco-based Global Exchange sued Nestlé USA, Archer Daniels Midland and Cargill on behalf of former child slaves. The initial 2005

complaint summary states in part: "Plaintiffs, aged 12 to 14 when first forced to work as child slaves, had to work 12- to 14-hour days with no pay. They often worked with guns pointed at them, and were given only the bare minimum of food scraps….Plaintiffs were locked in small rooms at night with other child slaves so they could not escape the plantations. They were whipped and beaten by the guards and overseers when the guards felt they were not working quickly or adequately."

"I tried to run away but I was caught … as punishment they cut my feet and I had to work for weeks while my wounds healed," one of the three boys says on videotape. All three are now adults; they moved to the city of Sikasso, Mali, after escaping slave-like conditions on plantations in Abobogou and the Region de Man in Côte d'Ivoire.[28]

The case was filed in California and was initially allowed to proceed, but in a 2021 ruling the Supreme Court declared that the plaintiffs needed to show that the abuse was due to decisions made in the U.S.[29]

U.S. social mobility as myth. Among the arguments used to defend the concentration of wealth found in the U.S. is one that claims that the wealthiest have arrived at their exalted position through meritorious struggle and that the American dream of millionaire status is available to all. There is, indeed, movement up and down the social ladder, but the numbers show that moving up from the bottom is quite difficult and that social mobility is greater in other countries. Consider that 40% of adults in the top (fifth) quintile remained there and most (63%) stayed above the middle, while only 4% at the top came from the bottom quintile. Those born into the bottom quintile often stayed there (43%) and 70% stayed below the middle; these figures are worse for Blacks.[30] Intergenerational mobility—when the child does better than the parent—is less in the U.S. than in other rich countries. Countries with stronger social safety nets than the U.S. have been found to have greater social mobility.[31] Another factor is the high-income inequality here: cross-country studies show that higher inequality is associated with a lower rate of upward mobility.[32] In addition, even if persons born into the bottom quintile move up a rung, they may be earning about the same income their parents did, because average weekly earnings have been falling in real terms since 1975 (average real weekly earnings in 2018 were lower than in 1975).[33] Families at the very top of the income distribution, however, have been making particularly large gains over the past several decades. Contributing to the problem, socioeconomic disparities translate into educational deficits, beginning with school readiness skills which emerge before kindergarten. Education does not eliminate early socioeconomic gaps, according to some studies.[34]

It is a myth, then, that everyone has a good shot at becoming rich. Millionaires seldom start from scratch. True, many of those whose wealth comes from businesses they own, or investments they have made, say that their success is due to their personal qualities and perhaps a bit of luck, not to inherited privilege. They are often disciplined and optimistic people. However, many if not most started their climb with economic advantages over the average American. Consider the facts behind the Forbes 400 list of the wealthiest Americans. Forbes states that 70% of the Forbes 400 "made their fortunes entirely from scratch." But most, as it turns out, came from families that were at the top of the top quintile. With a bit of digging, researchers found that more than a quarter of the Forbes 400 in 2012 inherited over $50 million. Another 11.5% came from households leaving inheritances of over $1 million and 22% came from upper-class families, with inheritances of up to $1 million. That is, at least 60% of the elite 400 started in the top quintile.[35]

Influence of the Wealthy on Policy and Public Opinion

The influence of the wealthy is not only economic and political, but also social, cultural, and ideological. As business owners and employers, the wealthy affect the lives and well-being of millions of employees and investors. The billions they spend on political campaigns and lobbying impact laws passed, regulations enacted and policies put in place by Congress and state legislatures. And there is more. Wealth, and the control of resources it brings, provide the means for the powerful to affect ways of thinking and decisions made throughout society. Their ability to sway public opinion can mobilize help for hurricane victims or spawn hate groups. Their outsized impact on our democracy is beyond troubling.

Political campaigns and lobbying. Every election year, billions are spent on federal elections; candidates in contested federal races spend hours every week drumming up financial support. Even in the nineteenth century, the corrupting influence of money on political campaigns and politicians was a growing concern. The first federal campaign finance law in the U.S., in fact, was passed in 1867.[36] Since then, several other laws aimed at campaign reform have gone into effect, apparently to little avail.

From 2012 to 2020, reported campaign spending on federal elections more than doubled to over $14 billion.[37]

Large donors to campaigns find their money well-spent. They can count on access to those candidates who are elected, which ensures that those politicians are acquainted with their concerns and preferences. Their hired lobbyists inform, wine and dine targeted politicians. In fact, hiring lobbyists is now de rigueur for big business, and the success of lobbyists in steering political agendas is documented. There is little doubt that "Protecting American interests" is at the top of national and international political priorities, and often these interests are conflated with the interests of big business. Taxes, expenditures, laws, and regulations are routinely assessed in light of American interests and of their impact on business.

Writing in 2015, journalist Lee Drutman noted that corporations and their trade associations spent more on lobbying ($2.8 billion) than the funding for the entire Senate ($860 million) and House ($1.18 billion) combined. Moreover, that $2.6 billion was about 34 times the total lobbying spending by all labor unions and groups representing public and consumer interests—organizations that could serve as a counter-weight to business pressures. Drutman (2015a) adds that "Of the 100 organizations that spend the most on lobbying, 95 consistently represent business…The financial and real estate industries alone spent $2.2 billion from 1999 to 2006 lobbying Washington, with good results in public policy." The defense industry is also a huge player in D.C., with U.S. military spending at $778 billion in 2020. The industry depends on federal contracts, and many regions of the country depend on the jobs it creates. Members of Congress themselves become lobbyists for military spending in their districts.[38]

Drutman notes that the top reason for lobbying was "to protect the company against changes in government policy" and that another important reason was the "need to improve the ability to compete by seeking favorable changes in government policy."

As the income gap has widened in recent decades, so has spending by the wealthy on lobbying. Prior to the 1970s, few corporations had their own lobbyists. But in 2019, there were almost 12,000 lobbyists in D.C. with expenditures of $3.5 billion.[39] Not surprisingly, lobbying has transformed how government functions.

Public opinion, the media, and think-tanks. In addition to their direct involvement with politicians and politics, the wealthy shape attitudes and public opinion on policy questions through their ownership and control of media outlets; through their role as business and community leaders; and through the think tanks, spokespersons, nonprofits and advertising they fund. Their influence extends to town councils and living rooms across the country, and paves the way for the successful implementation of policies they favor.

Public support for the decisions made by government and corporate leaders is critical to the effective exercise of authority and to stable social structures. Attitudes of the public toward on-going practices are fairly well established, but each new situation raises questions about those practices and new choices to be made. New technologies, changes in population, and other developments necessitate adjustments and realignments in organizations from families to the U.S. Congress. The pandemic has made this abundantly clear. Change means that political and business leaders are continually implementing new policies. Pointing the public toward choices being debated by leadership focuses attention on a narrow range of possibilities. In addition, the repetitious news coverage of a few facts and opinions reduces public interest in the mainstream news and shifts attention to more diverting entertainment or, recently, to conspiracy theories and the like. Alternative policy approaches favored by marginalized groups to issues of economic security, discrimination, and more are often not covered by the mainstream media and so do not even enter the debate. As a result, choices appear to the public to be limited to those favored by the Republican and Democratic leadership. This usually circumscribed agenda controls and limits national and state policy debates. Political developments in recent years have made a difference, however. The Biden agenda to Build Back Better marks a departure from the more limited programs that have been typical of the proposals of Congress and the White House. More on this in Chapter 6.

Ownership of the media+ Our opinions about world events, political candidates, and policy debates are formed by knowledge gained from many sources. We rely on information that comes from personal observation, from online and printed materials, from discussions and presentations, and from mass media outlets. Among these sources, information and opinions coming from media giants have the broadest dissemination. This gives the mass media considerable clout: widely shared opinions are necessarily based on information that is widely available—and that information is provided by the mass media.

Ownership of mass media conglomerates is highly concentrated in the hands of the wealthy. The numbers change over time, but the trend of increasing concentration is disturbing. Thirty years ago, media companies (films, video, and print apart from online) were owned by many different companies. Today, fewer than 10 companies dominate the market. The dominant media giants in late 2021 are Walt Disney (owner of 21st Century Fox, ABC, ESPN, Hulu, and more); Discovery/WarnerMedia (owner of CNN, HBO, Warner Brothers and more); Fox Corporation; Comcast (owner of NBC, Sky Group...), and ViacomCBS (owner of CBS, Paramount, MTV, Showtime...). As online news and opinion grow in importance, Netflix, Amazon, Google, and Apple are joining the more traditional forms of mass media with their streaming of news, documentaries, and the like. However, their news coverage is largely drawn from stories that originate in traditional media sources that have reporters in the field. Meta (Facebook rebranded), a social media giant, provides minimal editorial oversight, but its algorithms have a huge impact on what readers see.

Media giants are all conglomerates with holdings in industries that range from oil to real estate and local newspapers. Conflicts of interest between the public's right to know and the goals of these companies affect news coverage. FAIR, an organization devoted to Fairness & Accuracy In Reporting, cites cases in which media outlets killed stories because they undermined corporate interests; cases in which advertisers used their financial clout to squelch negative reports; and cases in which powerful businesses used the threat of expensive lawsuits to discourage legitimate complaints.[40]

"The institution of a free press in America is presently in a state of crisis greater than I have ever seen in my lifetime, and perhaps in any moment in this nation's history," said former television news anchor Dan Rather in 2018, and the situation since then has only worsened.[41] There are some independent national online news sources, but these are few in number. Research done by Titlemax found 23 such independent news sources, including the New York Times, Reuters, and AccuWeather.[42] These, however, are heavily influenced by their owners or largest shareholders, who are typically among the wealthy.

FAIR, as part of its media watchdog mission, has examined in depth the close ties between journalism and business, and the impact of these ties on information provided to the public. "With U.S. media outlets overwhelmingly owned by for-profit conglomerates and supported by corporate advertisers, independent journalism is compromised," says FAIR. "In this incestuous culture, 'news' is defined chiefly as the actions and statements of people in power. Reporters, dependent on 'access' and leaks provided by official sources, are too often unwilling to risk alienating these sources with truly critical coverage." Their critique of the press extends to National Public Broadcasting, which relies heavily on funding from Congress and shapes its news coverage in ways designed to avoid any adverse reaction from that body.[43] Big media companies routinely make large contributions to both major political parties, and, in turn, receive millions of dollars in payments for political ads. The owners and managers of major media outlets tend to be in the same income group as powerful politicians. Close ties between the media and politicians are one result and, not surprisingly, top news executives and reporters frequently socialize with top government officials.

Systematic Biases in Policy Research

Journalists and news commentators, besides shining a spotlight on the powerful, also present information from the studies that influence policymakers. These studies may be done by researchers in think-tanks, at universities, or in private consulting firms. These researchers present statistics and data related to policy questions facing lawmakers, and provide analysis and recommendations about alternative policy approaches. These studies, though, are not unbiased. There are liberal and conservative think-tanks; there are also consultants who are hired to provide information and analysis for a particular purpose; and there are nonpartisan research institutes that are funded by a variety of sources and that typically maintain good relationships with their major funders. These research institutes are generally funded by foundations, business groups, governmental bodies, nonprofits and well-to-do individuals.

Funding sources can have both direct and indirect influence on the studies done. Think-tanks that are liberal or conservative hire researchers who share the group's general perspective, and this perspective affects how a subject is approached. Nonpartisan research institutes are influenced by practical considerations: they want their research to result in policy recommendations that are politically viable or at least acceptable to a number of legislators—a goal usually shared by their donors. Consultants can be hired to flesh out and present a legislative proposal in the most favorable light possible rather than to objectively assess a proposal or economic trend.

The American Legislative Exchange Council (ALEC) is a think-tank-plus. This immensely influential organization brings together experts, legislators, and funders with a conservative bent. Among other work, the group develops model legislation that is regularly introduced in some form in state legislatures across the country.[44]

University researchers have more independence than paid consultants or think-tanks. They, however, are concerned with advancing their standing within their profession. This impacts their work choices; they usually focus on publishing in professional journals that require they make an original contribution to their fields—and research showing that a piece of legislation will have adverse impacts on society and the environment does not often require any original theoretical work. And with hundreds of bills before Congress and legislatures every year, few bills can be thoroughly analyzed. Policy studies for use by policymakers or by different interest groups is considered to be applied research and is usually funded by grants which come from the same sources that fund think-tanks.

Another significant source of bias in economic studies, apart from funding sources, is economic theory itself. We discuss this further in Chapter 5 where we take up economics and market ideology.

Statistics Reveal and Conceal

The growing economic inequality in the country is a topic that is affected by the kinds of biases we've just discussed. Research on the causes of the rising gap in incomes illustrates how statistics can be used selectively to tell different stories about which factors are the cause of the growing inequality. Krugman, a New York Times columnist and Nobel prize winner in economics, notes that, after an initial period during which rising inequality was not considered a problem, three theories emerged to explain the trend: globalization, technological change, and the superstar theory.[45] Both the globalization and technological change theories focus on trends in the demand for workers: fewer workers are needed in high-paying yet low-skill manufacturing jobs (many of which have gone overseas) and at the same time demand for high-skill workers is up in other industries due to new technologies. The result, it is claimed, has been a drop in the earnings of unskilled workers and increasing wages for higher-skill workers. The superstar theory notes that superstar athletes, actors, and musicians contribute to inequality as their pay soars. These arguments can account for some of the income gaps. But, as Krugman notes, they fail to explain the growing wage gaps within higher-skill workers, that is, the college-educated. Nor do they explain the skyrocketing of incomes of non-superstar CEOs, financial executives, surgeons, corporate lawyers, and others at the top.[46]

In examining inequality, research organizations with significant backing from labor unions, such as the Economic Policy Institute, probed income trends at disaggregated levels in depth. They looked closely at wage trends for different groupings of workers: blue-collar, white-collar, male, female, Black and White, and by educational levels. They were among the first to note the growing disparity among college-educated workers. As it turns out, real hourly wages declined for the bottom 70% of the college-educated workforce between 2000 and 2011, and the gap appears to be continuing. They also noted that the skyrocketing pay of CEOs relative to that of the average worker in their companies was not justified by the success of the CEO (in fact company performance had often worsened); to account for the excessive pay packages it was necessary to take changing corporate culture into consideration. And the falling percentage of workers in unions was also found to have explanatory power.[47]

Researchers, then, can often back up a particular storyline if they are selective about which variables they consider. In the case of inequality, a bit of digging showed that political and cultural factors are important and that levels of education are only one factor. The political clout of workers versus management can be difficult to measure, but some proxies can be used. Union membership numbers provide one indicator of labor's power, as does a company's reliance on temporary and part-time employees, on independent contractors, and/or on subcontractors. Trends in enforcement of federal workplace regulations can also be telling.

Organizations advocating for workers, women, the disabled, minority groups, and others have limited funds available to finance research, particularly in comparison to big business, foundations, and government agencies. They can only do a few in-depth studies annually. Better policy recommendations will result if we have more research done on behalf of such groups, especially on policies that are heavily favored by business, but that have severely negative consequences for many communities or workplaces. For example, how much should have been spent on bailing out people rather than on bailing out business in response to the Coronavirus pandemic?

Another problem is that highly relevant research often receives little coverage by the media; this is true when it comes to workers' demands.

Consider the periodic efforts to raise the minimum wage. Usually, articles and news reports cover "both sides." Business owners who oppose raising the minimum wage typically get equal time with proponents. The facts about minimum wage workers and the broad support among workers for a higher minimum get scant coverage. But data is readily available on such relevant information as the ages of minimum wage workers, the importance of the worker's income to their families and the percentage of them who are supporting children (about 30% of those working for the minimum wage in 2016 were supporting children).[48] The misinformation that is frequently heard in interviews is seldom corrected by the journalist. And the fact that the purchasing power of the minimum wage is worth less than in 1968, at $7.25 in mid-2022, may not come up.

Which Policy Is Best? For Whom? Standpoints and the Role of the Press

Which policy is best? The answer depends on asking first: best for whom? It is important that the public and policymakers hear the information supporting the viewpoints of each group that will be strongly affected by the policy. Objectivity demands this and only those groups impacted can confirm that their viewpoints and best interests are being presented accurately. Before a group can arrive at a policy position, it must have the information it needs and take the time required to digest and debate alternative responses. This means the group must be self-organized.

Philosophers agree that even objective observers cannot have an unbiased opinion. We are all "biased" by the knowledge and opinions we have. The most we can hope to get from our information sources are reports and opinions that are based on the best facts and theories available, ones that take into account the standpoints of the different groups involved (Fraser 1989). And we know that new information and new scientific findings will continually modify our understandings.

The process used to identify a group's true/best interests regarding a policy is crucial. The self-organization of a group is necessary, for one thing. Without some minimal cohesion, a group cannot hold the discussions needed to determine its true interests. True interest theory lays out some of the pre-requisites for recognizing and articulating true interests (Hamilton 1999). First, an attempt at objective assessment of that group's well-being from different positions is required, using information that combines external and internal valuations. This process must also be followed in assessing the likely impact of a policy change on the group.

Needs often remain unaddressed for years, as in the case of worker safety. Workers may lack accepted channels for meeting and discussing needs, let alone a means of redress. When a power imbalance exists, as at the workplace, greater democratization offers a partial remedy. Public conversations, by enabling groups and individuals to better identify, refine, and decide upon their interests, are also valuable to this process. An informed public improves the functioning of democracy; and the media has a critical role to play in seeking out and presenting the standpoints of all groups impacted by a proposal.

While it may be difficult to determine the best or true interests of a group, at the least, findings of a research group with one perspective should be balanced with the findings of researchers with other perspectives on important issues. Congress could provide needed funding on issues impacting a large part of society, e.g. what would be the costs and benefits to society of providing childcare/early childhood education?

Shaping Opinions and Policy

The efforts the wealthy put into lobbying Congress, funding political campaigns, financing research, and overseeing media giants all pay off. These investments facilitate the smooth functioning of society, under laws and policies that have been put into place with the approval or acceptance of the powerful. The information shaping opinion about laws and policies is largely provided by the mass media and by articles, research, and even tweets funded or authored by the well-to-do. The extent of public acceptance of, and compliance with, government laws and policies depends on such messaging.

Issues debated in the media in 2022 include questions related to COVID-19, the war in Ukraine, privacy and security, immigration policies, access to abortion, and much more. New policies that are implemented change not only how things are done but also our understanding of how things should be done and why. Media coverage of top-down change paves the way for a smooth implementation process. If the media is divided, public reaction will be as well.

As an example of how even major changes can be accomplished with little blow-back, consider the establishment of the Department of Homeland Security (DHS) after 9/11. DHS ushered in new practices affecting security protocols across the country with little adverse public reaction. Control over how the public perceives changes in practices and policies strengthens the position of authorities. Change from the bottom-up is not as easily accomplished.

Concluding Remarks

The wealthy differ from the rest of us in terms of economic security, friendships and lifestyles, political preferences and political clout, ownership and control of the media and other businesses, and more. Congress is swayed by business lobbyists, while public interest groups and voters have been shown to have much less influence. We examine the shared beliefs and institutionalized practices that uphold the domination of society by the wealthy in the next chapter, how that domination has affected our democracy, and how new ways of thinking can turn things around.

The moral challenges of today are the equals of those of earlier times. But the struggles to ensure that right prevails now hold the dire possibility that lost battles will leave in their wake a devastated planet with diminished prospects for coming generations of humanity. These battles demand the same qualities as ever: courage, strength, self-sacrifice, and a belief in the power of good. Heart-breakingly difficult tasks must be faced; obstacles may entangle, or strangle, emerging hopes. But only a transformation of our societies can provide the foundation for a community of humanity whose leaders are dedicated to the collective good. This shift in direction is vital; it demands an end to the practices that protect and privilege the powerful few who are jeopardizing our common future. The belief that this is possible is a belief that can make this possibility a reality.

Notes

1. https://www.pewresearch.org/global/database/indicator/6/, https://www.pewresearch.org/politics/2019/09/19/why-americans-dont-fully-trust-many-who-hold-positions-of-power-and-responsibility/, https://www.pewresearch.org/politics/2020/09/14/americans-views-of-government-low-trust-but-some-positive-performance-ratings/, and https://ourworldindata.org/grapher/public-trust-in-government.
2. https://news.gallup.com/poll/1597/confidence-institutions.aspx. Confidence was high in the military (73%), small business (68%) and the police (53%), as of 2019. Churches and banks rated lower.

3. https://www.politico.com/magazine/story/2017/07/20/18-confusing-contradictory-and-just-plain-kooky-things-donald-trump-has-said-about-health-care-215402/.
4. http://nymag.com/intelligencer/2018/05/trump-broken-every-populist-campaign-promise-prescription-drugs-infrastructure.html.
5. "America is born of change.... the single-most powerful word in our democracy is the word 'We.' 'We The People.' 'We Shall Overcome.' 'Yes We Can.'" Obama, 2015 at Selma. https://chicago.suntimes.com/columnists/2021/3/7/22318510/obamas-2015-selma-speech-excerpts-to-be-chisled-on-obama-presidential-center-facade.
6. https://www.politifact.com/truth-o-meter/promises/biden-promise-tracker/?ruling=true.
7. Wolff (2021: 50, Table 4).
8. Wolff (2021: 56, Table 10). See also Saez and Zucman (2020).
9. https://www.forbes.com/sites/forbespr/2020/09/08/forbes-releases-39th-annual-forbes-400-ranking-of-the-richest-americans/?sh=483b66bf335b.
10. https://www.federalreserve.gov/econres/scf/dataviz/scf/chart/#series:Net_Worth;demographic:inccat;population:all;units:median.
11. https://press.careerbuilder.com/2017-08-24-Living-Paycheck-to-Paycheck-is-a-Way-of-Life-for-Majority-of-U-S-Workers-According-to-New-CareerBuilder-Survey, https://psc-cuny.org/clarion/march-2012/how-business-elites-looted-private-sector-pensions.
12. https://www.federalreserve.gov/newsevents/pressreleases/other20180522a.htm. https://www.federalreserve.gov/newsevents/pressreleases/other20180522a.htm.
13. https://www.census.gov/data/tables/time-series/demo/families/households.html. There were 128,579 thousand households in U.S. 2019. The net worth figure is from Wolff (2021).
14. Wolff (2021, Table 4).
15. https://www.irs.gov/pub/irs-soi/soi-a-ints-id1801.pdf.
16. https://www.opensecrets.org/news/2020/04/majority-of-lawmakers-millionaires/. Top 100 (written 7/24/2014). This chart shows the average yearly percentage change in net worth of the 100 congressional incumbents whose calculated net worth divided by the number of years studied was the highest.

    - The average increase in net worth in the Top 100 was 114% per year.
    - Of the "Top 100", 56 are Republicans, 43 are Democrats and one is an Independent. In total, the study looks at 320 Republicans, 296 Democrats and two Independents. https://ballotpedia.org/Changes_in_Net_Worth_of_U.S._Senators_and_Representatives_(Personal_Gain_Index).

Data on net worth of members of Congress is available at http://www.opensecrets.org/personal-finances/top-net-worth. Disclosure of net worth within a range is required by law.

17. https://sgp.fas.org/crs/misc/97-1011.pdf. A Congressional Research Service publication. Additional income is permitted, but is less than $30,000.
18. Page et al. (2013: 66). See also Page et al. (2019). https://fivethirtyeight.com/features/everyone-knows-money-influences-politics-except-scientists/.
19. https://www.people-press.org/2016/04/26/a-wider-ideological-gap-between-more-and-less-educated-adults/.
20. Gilens (2012: 119). Fewer than 20% of the wealthy favor redistributing wealth through higher taxes on the wealthy, although most others polled were in favor, according to Page et al. (2013: 63). The Occupy Wall Street (OWS) movement of 2011–2012 grew out of a recognition that even the basic requirement of the American dream—a job with an income sufficient to support a middle-class standard of living—was becoming unattainable for many who played according to the rules. OWS organized street protests against the greed of big money (the 1%) and its indifference to the plight of hard-working Americans.
21. The following says in brief that rich Democrats are more economically conservative than typical Democrats and rich Republicans more socially liberal e.g. about gay rights. https://www.huffpost.com/entry/new-evidence-that-the-ric_b_7153396, https://www.people-press.org/2016/04/26/a-wider-ideological-gap-between-more-and-less-educated-adults/. Pew Research: Support for legal abortion increases with education: https://www.pewforum.org/fact-sheet/public-opinion-on-abortion/.
22. Politicians make decisions about government expenditures, taxes, laws and policies. Their decisions represent compromises reached within the bodies to which they are elected. Those in leadership positions such as chairmanships are under added pressure to adhere to the party line, to carry out the directives they have helped set for their party and, to be effective, are well aware of who favors or opposes particular policies and the extent to which compromise is possible while still satisfying party goals.
23. Office of the Comptroller of the Currency, *Quarterly Report on Bank Trading and Derivatives Activities.*
24. A former investment manager at a big firm discusses the ability of the wealthy to avoid taxation: http://whorulesamerica.net/power/investment_manager_2014.html.
25. https://www.Nestlé.pk/featuredstories/fla-report-cocoa_Nestlé.
26. https://www.theatlantic.com/health/archive/2012/03/do-children-harvest-your-food/254853/. In 2012, 500,000 children picked about 1/4 of food harvested in U.S., including some as young as six. See

also Association of Farmworker Opportunity Programs report: https://afop.org/wp-content/uploads/2010/07/Children-in-the-Fields-Report-2007.pdf, http://nfwm.org/resources/children-in-the-fields/. According to the law, children must be over the age of 12 to work in agriculture and must have their parents' permission, but in fact these restrictions are not well enforced. Most migrant teens working in fields are living by themselves. Health impacts—physical and emotional—are great. The children are handling pesticides and dangerous equipment—most (70%) tractor/vehicle accidents in field in 2002 occurred when driven by children. https://www.hrw.org/news/2019/11/13/children-working-terrifying-conditions-us-agriculture.

27. https://www.hrw.org/reports/crd0510_brochure_low.pdf.
28. http://www.corpwatch.org/article/chocolate-slavery-case-against-Nestlé-allowed-proceed.
29. In June 2021, The Supreme Court reversed a lower-court ruling that had allowed six men to sue Nestlé USA and Cargill over claims they were trafficked as child slaves to farms in the West African nation of Ivory Coast that supply cocoa to the two companies. https://www.cnbc.com/2021/06/17/supreme-court-rules-in-favor-of-Nestlé-in-child-slavery-case.html. The U.S. Supreme Court decided that the 1789 Alien Tort Statue applies only if the abuse was due to decisions taken in the U.S.
30. https://www.pewtrusts.org/~/media/legacy/uploadedfiles/pcs_assets/2012/pursuingamericandreampdf.pdf. See Figure 3, page 6, of *Pursuing the American Dream*.
31. https://library.cqpress.com/cqresearcher/document.php?id=cqrglobal2010080000, on Europe's social mobility.
32. Russell Sage Foundation (2016: 6).
33. Council of Economic Advisers (2021, Appendix, Table B-30).
34. Russell Sage Foundation (2016: 1).
35. http://www.faireconomy.org/bornonthirdbase2012, https://inequality.org/research/selfmade-myth-hallucinating-rich/.
36. https://www.washingtonpost.com/news/the-fix/wp/2014/04/03/a-history-of-campaign-finance-reform-from-george-washington-to-shaun-mccutcheon/.
37. https://www.opensecrets.org/campaign-expenditures?cycle=2020, https://www.fec.gov/data/spending-bythenumbers/. Data on campaign spending at https://www.opensecrets.org/overview/cost.php. Lee Drutman (2015b) notes that very few companies had their own Washington lobbyists prior to the 1970s.
38. U.S. military spending was $778 billion in 2020, more than the amount spent by the next nine top-spending countries combined. China spent

about $252 billion and Russia, $62 billion. The size of the military-industrial complex makes movement toward peace more difficult. The Government Accountability Office regularly reports on grossly inefficient expenditures in individual contracts and sometimes in entire weapons systems. https://worldpopulationreview.com/country-rankings/military-spending-by-country.

39. https://www.cnbc.com/2020/10/05/q1-lobbying-spend-was-record-938-million-but-lobbyists-decry-stereotype.html.
40. https://www.thenation.com/article/environment/media-fossil-fuel-ads/, https://www.forbes.com/sites/stephaniedenning/2018/09/19/why-jeff-bezos-bought-the-washington-post/?sh=4677254a3aab.

    Which company owns which is always changing. The following information is already out of data but the big six media giants controlled most of cable recently and now may control more. Their income is in the hundreds of billions.These giants can be broken into sub companies; for example, News-Corp included Fox, Wall Street Journal, and New York Post. Disney included ABC, ESPN, Pixar, Miramar, and Marvel Studios. Viacom included MTV, Nick Jr, BET, CMT, and Paramount Pictures. Time Warner cable included CNN, HBO, Time, and Warner Bros. CBS (merged with Viacom in 2019, forming ViacomCBS) included Showtime, Smithsonian channel, NFL.com, Jeopardy, and 60 minutes. Data can be updated using carefully selected sources including some citations on Wikipedia for corporate media. https://www.businesswire.com/news/home/20191119005997/en/New-Media-Gannett-Complete-Merger-Creating-Leading.

    Ashley Lutz (14 June 2012). "These 6 Corporations Control 90% of the Media in America". *Business Insider*. 2012.

    Judith Hollar (February 5, 2020) in https://fair.org/home/journalist-politician-corporate-coziness-on-display-at-snowflake-garden-brunch/: "The Snowflake Garden Brunch …where lobbyists throw parties for journalists so they can schmooze with the politicians they cover—sponsored by the same corporations that fund (and sometimes own) the media outlets the journalists work for."

    See also https://www.pewresearch.org/topics/media-ownership/.
41. Dan Rather (August 16, 2018) in https://www.theatlantic.com/ideas/archive/2018/08/why-a-free-press-matters/567676/.
42. https://www.titlemax.com/discovery-center/lifestyle/who-owns-your-news-the-top-100-digital-news-outlets-and-their-ownership/.
43. See "What's wrong with the news" at fair.org. (https://fair.org/about-fair/whats-wrong-with-the-news/). Joseph Atkins (2008), *Covering for the Bosses*, studies bias in the Southern press.
44. https://alec.org/about/. "Bills based on ALEC models were introduced nearly 2,900 times, in all 50 states and the U.S. Congress, from 2010

through 2018, with more than 600 becoming law, the USA TODAY and *Arizona Republic* analysis found." https://www.usatoday.com/story/news/investigations/2019/04/03/alec-american-legislative-exchange-council-model-bills-republican-conservative-devos-gingrich/3162357002/.

45. Krugman, Paul. "The Dangers of Extreme Wealth Concentration", 3/28/14. *New York Times* and https://www.nytimes.com/2002/10/20/magazine/for-richer.html.
46. https://www.brookings.edu/research/make-elites-compete-why-the-1-earn-so-much-and-what-to-do-about-it/.
47. https://www.epi.org/publication/10-year-decline-wages-college-graduates/, https://www.epi.org/data/#?subject=wage-education. Those with a bachelor's but no higher received same wage over the 2000–2014 period: https://www.bls.gov/opub/ted/2018/weekly-earnings-by-educational-attainment-in-second-quarter-2018.htm, https://fas.org/sgp/crs/misc/R45090.pdf, p. 12 modest wage growth for each education level, except for a drop for those with some college…

    From Page, Bartels and Seawright (2013): "We report the results of a pilot study of the political views and activities of the top 1 percent or so of U.S. wealth-holders. We find that they are extremely active politically and that they are much more conservative than the American public as a whole with respect to important policies concerning taxation, economic regulation, and especially social welfare programs…..the top one-tenth of 1 percent of wealth-holders (people with $40 million or more in net worth) may tend to hold still more conservative views that are even more distinct from those of the general public." They mention Jacob Hacker and Paul Pierson on how Washington has "made the rich richer."
48. https://www.epi.org/publication/15-by-2024-would-lift-wages-for-41-million/.

References

Callahan, David, and J. M. Cha. 2013. *Stacked Deck: How the Dominance of the Affluent.* Report for Demos. https://www.demos.org/research/stacked-deck-how-dominance-politics-affluent-business-undermines-economic-mobility-america.

Council of Economic Advisers. 2021. *Economic Report of the President 2021.* Statistical Appendix, Table B-30.

Domhoff, G. William. 2020. *Who Rules America.* Online report continually updated. University of California at Santa Cruz. https://whorulesamerica.ucsc.edu/.

Drutman, Lee. 2015a. Better Way to Rein In Lobbying. *New York Times*, April 24.

———. 2015b. How Corporate Lobbyists Conquered American Democracy. *The Atlantic*, April. https://www.theatlantic.com/business/archive/2015/04/how-corporate-lobbyists-conquered-american-democracy/390822/

Fraser, Nancy. 1989. *Unruly Practices*. Minneapolis: University of Minnesota Press.

Freeland, Chrystia. 2013. *Plutocrats*. New York: Penguin Books.

Government Accountability Office. 2008. Media Ownership: Economic Factors Influence the Number of Media Outlets in Local Markets. GAO-08-383.

Gilens, Martin. 2012. *Affluence and Influence: Economic Inequality and Political Power in America*. Princeton, NJ: Princeton University Press.

Gilens, Martin, and Benjamin Page. 2014. Testing Theories of American Politics: Elites, Interest Groups, and Average Citizens. Published online by Cambridge University Press, September 18. https://www.cambridge.org/core/journals/perspectives-on-politics/article/testing-theories-of-american-politics-elites-interest-groups-and-average-citizens/62327F513959D0A304D4893B382B992B (Also in *Perspectives on Politics* 12 (3): 564–581).

Hamilton, Lawrence. 1999. A Theory of True Interests. *Government and Opposition* 34 (4): 516–546.

Page, Benjamin, Larry Bartels, and Jason Seawright. 2013. Democracy and the Policy Preferences of Wealthy Americans, January. A Russell Sage Foundation Report. https://www.russellsage.org/research/reports/wealthy-americans-political-preferences.

Page, Benjamin, J. Seawright, and M. Lacombe. 2019. *Billionaires and Stealth Politics*. Chicago: University of Chicago Press.

Russell Sage Foundation. 2016. What We Know About Economic Inequality and Social Mobility. 2016 Report at: https://www.russellsage.org/what-we-know-about-economic-inequality-and-social-mobility-united-states.

Saez, Emmanuel, and G. Zucman. 2020. *Trends in U.S. Income and Wealth Inequality*. National Bureau of Economic Research. Working Paper 27921, October.

U.S. Treasury, Office of the Comptroller of the Currency. Quarterly Reports 2007–2022. *Quarterly Report on Bank Trading and Derivatives Activities*. https://www.occ.gov/publications-and-resources/publications/quarterly-report-on-bank-trading-and-derivatives-activities.

Wolff, Edward. 2021. Household Wealth Trends in the United States, 1962 to 2019. Working Paper 28383. Cambridge, MA: National Bureau of Economic Research. http://www.nber.org/papers/w28383.

CHAPTER 4

Fixing Our Broken Democracy

As we saw in Chapter 3, the preferences of the average American have a near-zero impact upon public policy, according to a statistical study by Gilens and Page (2014: 575). Other studies have similarly found a large gap between the preferences of the public and policies enacted by Congress (Page et al. 2013). At the same time, business lobbyists are quite effective in influencing policy, according to these studies. The problem has not gone unnoticed by the public: most Americans believe that big business and lobbyists have too much influence over politicians. One survey found that an impressively high 84% of those polled believed that money has too much influence in political campaigns.[1] Another poll found that 72% of respondents thought social media companies had too much political influence.[2] Most Americans also agree that Congress pays more attention to donors than to constituents.[3]

Low levels of confidence in several aspects of our democracy are confirmed by other surveys of the voting population. Key to the lack of confidence in government is the problem of poor representation in legislative bodies, and this is a problem that has worsened over the past decade. The U.S. has been ranked as a "flawed" rather than a "full" democracy since 2015 in the comprehensive rating of the world's nations done annually by a leading U.K. research group. The Economic Intelligence Unit (EIU) looks at several factors in determining its rating of the world's democracies. One area where the U.S. falls short is in accountability to

M. T. Hill, *Building a Solidarity Society*,
https://doi.org/10.1007/978-3-031-07349-6_4

voters: "If democratically based decisions cannot be or are not implemented, then the concept of democracy is not very meaningful" (EIU 2020: 56). Several statistical measures address this point. Also among the factors contributing to the drop in the U.S. rating was a fall in confidence in the functioning of public institutions. Low voter turnout (below 70% even in 2020) in comparison to other countries has also pulled down the U.S. rating.

Belief in the strength of the U.S. democracy is indeed eroding, and there is a good reason. As we noted in Chapter 3, even such a basic value as opposition to slavery and human trafficking is weak in Congress: a bill that would have prevented federal contracts from going to firms that engaged in human trafficking did not make it out of committee.

Our Country: Plutocracy and Democracy

We are ready for change. We are angry that the wealthy dominate policy-making to the extent that they do. Their interventions in elections, media programming, policy research, and the legislative process make the U.S. as much a plutocracy as it is a democracy. A **plutocracy**, by definition, is a society governed by the wealthy. Freeland (2013) makes clear the social prerogatives of plutocrats. If the U.S. is, in effect, a plutocracy and if there is broad agreement that the power of our plutocrats is too great, why is it that this situation persists year after year, given the democratic tools at our disposal? A short answer is that the power elite have a hegemonic hold. This means that they exert strong political, cultural, ideological, and economic influence over society. They, in fact, dominate social institutions. Their **hegemony** has been built up only slowly over decades or even centuries.[4] It rests on the intertwining of the institutions, identities, and ideologies that bind a society together. The stronger these interlocking bonds are, the more stable the society and the more formidable the hold of the prevailing hegemony.

Ending the domination of the wealthy, then, is not simply a matter of electing politicians who support campaign reform or other measures that would deepen democracy. The hegemony of the wealthy must be challenged by alternative views of society and of how democracy should work. Doing so will bring pervasive change to our understanding of social and economic realities. Criteria used to define poverty, unfair labor practices, and more, for example, will change.[5] Today, the hegemonic hold of the wealthy is fraying; the ideologies supporting their authority have weakened. Divisive ideologies and conspiracy theories are growing in popularity, but so are demands for living wages, access to health care, and policies that take on climate change. Those discontented with present policies likely share the belief that they and their families deserve a better shot at the "American dream" than they now have. Overcoming political divisions is key to building populist coalitions that can, in fact, bring the "American dream" closer to reality for all. In this chapter, we provide a theoretical framework for understanding the hegemony of the elite and for shaping new ideologies that can unite people in efforts for transformative change.

Ideologies support the hegemonic hold of the plutocrats. The ruling elite either head society's major institutions, or they have veto power, effectively, over who will head those institutions. This enables them to ensure that the laws, policies, and practices that guide society align with their interests, broadly defined. (This is not to deny that there are always internal disagreements among the elite. Areas of strong agreement, such as the protection of business rights, were laid out in Chapter 3.) The elite are able to make decisions that affect businesses, government, and the nation at large. This power is possible because people accept their authority, and critical to this acceptance are the shared beliefs that support a nation's institutions—the national ideology.

The national **ideology** consists of the shared beliefs that are used to explain and guide national practices and policies. A belief in human rights, for example, is part of the national ideology. This belief, as well as beliefs in justice and liberty, can be rather vague beliefs that are shared in some form by conservatives, liberals, and most citizens. All 192 member countries of the U.N. have signed on to the Universal Declaration of Human Rights which includes among its values life, liberty, security of person, freedom of speech, universal suffrage with secret ballot, fair public trials and equality. Most countries also profess a belief in fighting climate change: 174 countries signed the Paris Accord on Climate Change. The laws and practices of these countries, however, vary substantially in their protection of human rights and in their effectiveness in fighting climate change. Debates about policies within countries usually entail heated exchanges about any particular course of action, given different understandings of basic ideological beliefs and policy options.

To speak of "the national ideology" is not, in fact, accurate. Every nation has a mix of ideologies: religious, political, social, and others. Each of these ideologies presents a simplified description of one aspect of society, usually using stereotypes about people that legitimize and support shared beliefs about how to treat different groups. When ideologies clash, the one that is most in line with national practices is said to be dominant and so to have more explanatory power. Dominant national ideologies, as they justify the status quo, also justify existing inequalities and injustices.

Political scientists usually discuss dominant and competing ideologies rather than a single national belief system. Take human rights: there are competing beliefs, not just one, about the right to health care, about women's rights, and about the rights of gun owners. More on this later.

Democracies reflect the ideological beliefs of their founders, and the strengths and limitations of those beliefs. Wealthy landowners were the founders of the U.S. and for many years only White, male property owners could vote. As one political scientist put it, "One would not err much in thinking that the strategic problem of founders, pretty much everywhere, was how to construct representative government for the rich while protecting it from the poor" (Przeworski 2010: 162). But Przeworski (2010: 169) also adds: "The fact that those endowed with disproportionate economic, ideological, and organizational resources use them successfully to defend their privileges does not render futile the struggle for economic and social equality." And struggles for equality have in fact changed attitudes, laws, and the prevailing ideologies of the U.S.

Shared Understandings: The Cement of Society

To understand how changes occur in ideologies, and in hegemonies, we must first acknowledge their deep hold on us. As we become part of a society, we absorb its shared understandings and function within the meaning system it has developed over time.[6] We learn how things are done, and some of the reasons why, through a socialization process that begins in childhood. Gradually, messages about what is socially acceptable behavior shape our routine response to different situations and to different groups of people. Internalized understandings about society enable us to survive and even thrive. Our identities emerge based on the understandings and behaviors we use to navigate life, in families and firms; in churches and in sports; and in the many social activities of our lives. We share many beliefs with our fellow citizens, about democracy, freedom of speech, human rights, and more. We associate positive values with certain kinds of behavior and are invested in protecting the values in which we believe. We are outraged when those with responsibilities fail to carry them out and people suffer as a result. We demand appropriate punishment for such misbehavior.

The status quo is upheld by our shared beliefs and understandings. It provides a framework ensuring that necessary tasks are done. As we perform routine tasks, like purchasing products, paying taxes, and caring for our children and family members, we contribute to society. We make distinctive contributions through our work. In brief, we enable the functioning of society as it stands. Like it or not, we are part of the status quo. All the wide variety of activities in a society, taken in total, have the effect of upholding the status quo, albeit with some change over time.[7]

But current practices also have the undesirable effects of creating and exacerbating inequalities and environmental crises. We may be passive in the face of unfair or environmentally degrading practices at work or at home; or we may oppose or support them. Together, we make social systems work as they do; as individuals, however, we can do little to change them. Working with others, though, we can start moving society in new directions. There is hope, then, that collectively we can deepen democracy, address climate change, and move toward solving other urgent problems. But the status quo is highly resistant to change and it is important to recognize the challenges involved in changing ideologies and established institutional practices.

Ideology: Gaps Between Ideals and Realities

Changing shared understandings is the first step toward changing established ways of doing things. Openings for re-examining accepted practices come when there is widespread recognition of a disconnect between a shared ideal and actual behaviors. For example, U.S. political ideologies hold up the ideals of liberty and justice for all, but at times the gap between this ideal and actual behavior sparks outrage and efforts for change. This was the case when George Floyd was killed by a police officer. After his death, efforts to improve attitudes and practices of police departments across the country intensified. George Floyd's death changed the attitude toward police departments and the ideologies that have guided them. Police chiefs and mayors are re-examining their modus operandi and making changes. This is a step toward meaningful reforms.

"Liberty and justice for all" is only one national ideal; every system in a society has ideals and ideologies. Justice is the guiding ideal of judicial systems; democracy, of the political system; love of neighbor and of God, the ideal of religions; enlightenment, the goal of educational systems; efficiency and productivity, ideals of the economy; and so on. But, as we know, actual practices of these systems fall short of the ideals they hold. Consider the economy and the ideal of efficiency. In theory, economic efficiency, and fairness, depend on equal pay for equal work. Ideological reasons are given to explain why women and people of color are paid less than White men on average for doing the same job with the same qualifications (see below). It is claimed that wages are determined by the market and the guidance of the market must be followed if a business is to be competitive. This is part of market ideology. There are many examples of inequalities and injustices that are justified by various ideological explanations.

Let's look briefly at the ideals and ideologies guiding the Founding Fathers of the U.S. The Bill of Rights and the Constitution express many of the ideals guiding the founding of the U.S. The Constitution states as its goal: "to establish justice.... promote the general Welfare, and secure the Blessings of Liberty to ourselves and our Posterity." The Bill of Rights lists, in part, the right to the free exercise of religion; "...freedom of speech, ... (of) the press; ... the right of the people peaceably to assemble, and to petition the Government for a redress of grievances." The Declaration of Independence lauds the rights to life, liberty, and the pursuit of happiness. Abraham Lincoln's famous Gettysburg Speech holds up the ideal of a government of, by and for the people. Taken together, this system of ideas and ideals forms part of the dominant national ideology.

But these ideals, which are often cited as guiding principles of the nation, do little to explain or justify how institutions actually function at present, or how they have functioned in the past. It is not hard to see that the positive ideological messages from early America do not hold up well in light of historical realities. Slavery, the denial of voting and property rights to women, men's control over women's bodies, and other injustices were national practices when the Declaration of Independence was signed and when the Constitution was adopted. These unjust practices were justified at the time by ideological messages claiming that Blacks, women, Asians, Indigenous peoples, immigrants, and many others were inferior to well-to-do propertied White men, and so unworthy of the same rights.

A look at religious beliefs clarifies the impact ideologies have. Every established religion is an ideology, a way of understanding reality, one that provides values and guidance. Religion affects behavior, at times motivating good works and missionary zeal, and at other times motivating persecution, ostracism, and holy wars. It is clear that beliefs associated with religion can be positive or negative; and leadership has a major influence. Beliefs apart from those embedded in religious ideals must be taken into account in order to explain what persons of a given religious faith, in fact, do. People are subject to cognitive dissonance—they hold contradictory beliefs simultaneously. They usually explain why they are doing something that apparently contradicts their religious beliefs by citing another belief. In general, people will hold to a rather idealistic ideology, such as a belief in equality, while at the same time also holding another that conveniently rationalizes why they act contrary to that ideal. As to equality: why shouldn't children have equal access to a good education? Or, to draw on an earlier example, why shouldn't women get equal pay for equal work? Oppressive ideologies explain why; they justify privilege and poverty, discrimination, and destruction of the environment.

Although beliefs alone do not account for behavior, beliefs in conjunction with behavior provide insight into human motivation. And understanding motivation is key to motivating the desire for change. In the case of national ideologies, belief in democracy, justice, capitalism, and individual freedom does not alone explain the content of laws adopted or the practices of government agencies. Laws prohibiting free Blacks from holding certain jobs and laws denying women property rights were justified by beliefs prevailing at the time they were adopted. In the case of the Founding Fathers of the U.S., besides their belief in "equal rights," they also believed that slavery was acceptable and that women did not merit access to the same rights as men. The dominant national ideologies of the time, then, included racist and sexist beliefs rationalizing oppressive practices as well as beliefs in progressive ideals. Oppressive ideologies are inevitable when there is oppression. The benefits to the Founding Fathers from their oppressive practices made it difficult for them to see the injustices involved in their treatment of others. It is a common human failing. Motivating people to change their oppressive behaviors can begin after they acknowledge the perceived benefits to themselves from that oppression.

Oppressive Ideologies: A Critical Obstacle to Unity

The ideologies of a country set out both positive ideals and rationalizations of behaviors that deviate from those ideals. In the U.S., there is widespread belief in justice, democracy, equal opportunity, freedom of speech, and more. There are also beliefs that elevate one group over another: racism, sexism, homophobia, nationalism, imperialism, and prejudice due to religious differences, among others. The abuse of the environment and its flora and fauna is also explained away by a range of beliefs.

Oppressive ideologies explain that marginalized groups doing less desirable work at low pay are different from the mainstream—they are "other." These ideologies claim that a group is suffering economically, physically, or in other ways because of qualities inherent to the group, not because of the practices of more dominant groups. The more dominant groups, though, are the ones with greater control of the circumstances facing the oppressed. The powerful dominate the public and private sectors, and have the deciding say in setting budgets and policies that determine educational opportunities, employment practices, and more. Their voices are respected. The fact that dominant groups derive psychological, social, and economic benefits from the subordinate status of others is significant. Ideological arguments by dominant groups that are conveniently self-serving should be regarded with suspicion.[8] For example, are poor Black teenagers who get pregnant the cause of poverty in Mississippi (as I have heard claimed), or do those in power have more control over the state's future through the decisions they make regarding investments, employment, taxes, and government spending?

Common practices are often considered to be "just common sense," but many are actually oppressive. For a simple example, it is "common sense" that women should be the ones expected to cook and bring the food for a communal potluck dinner—and also should be the ones in charge of cleaning up afterward. Is it also just common sense that women should be in charge of household tasks and the primary caregivers for children and elderly parents? In the U.S. and other countries, women typically spend more than twice as much time on household chores as their husbands, regardless of whether or not they have paid employment.[9] Sexist behavior is a norm widely accepted by both men and women.

Similarly, people of color are often hired for the least desirable jobs, at low pay with no career ladder. They often must have two or more jobs to support their families. A Black man in the U.S. in 2015 earned $0.78 for every dollar a White man of similar background earned.[10] Most of this gap is likely explained by discriminatory hiring and employment practices. Employers typically defend these hiring practices and also oppose policies that would improve pay, working conditions, or career ladders for those in low-paying jobs.

Women are often paid less than White men for performing the same or equivalent jobs.[11] The pay gap by occupation in the U.S. ranges from 10% in lower-paying occupations to 24% in higher-paying ones; it is greater for Black and Latino women. This pay differential is projected to add up to over $1 million over the course of a lifetime for college-educated millennial women. This failure to adhere to the principle of equal pay for equal work is justified by employers who cite market ideology. Wages are determined by the market, they claim, and they simply want to remain competitive. Market ideology, as we shall see, can justify many unjust but profitable practices.[12]

As the killing of George Floyd illustrates, the consequences of racism are not limited to gaps in pay or unequal access to housing and educational opportunities. Sexism and homophobia also result in sexual harassment and violence. Most women experience sexual harassment during their lives. Over 1000 women per year in the U.S. are killed by intimate partners, and thousands more are raped.[13] Violence against the LGBTQ+ community is also well-documented. LGBT people are much more likely to be victims of rape, sexual assault, and aggravated or simple assault than others, according to a study by the Williams Institute at UCLA.[14]

The rights of employees at the workplace leave much to be desired. There is no freedom of assembly or freedom of speech while on the job. But, in addition, workers have lost their jobs due to opinions expressed outside of the workplace or for frivolous reasons. Unlike other countries, the U.S. allows the firing of employees without cause. The Employee Bill of Rights suggests some of the protections that would benefit workers.[15]

Besides the harm done to those victimized by oppressive beliefs, these beliefs prevent the unity needed for working together effectively toward shared goals such as improved economic opportunities. The labor movement, the women's movement, and other loose coalitions have often been plagued by internal splits based on differences of race, gender, or other characteristics. Environmental groups that are largely White may not be sensitive to issues of racial justice. Labor unions have at times been slow to include women and Blacks as full, equal members—some still fail to take a strong stand for women's reproductive rights or police reform. Such divisions are welcomed by those whose power and control would be weakened by unified opposition.

Wealth: The "Royal Blood" of Capitalism

Capitalism fosters oppressive ideologies concerning workers and the nature of human relationships. Low pay, poor working conditions, and lack of voice in the decisions of the firm are explained by ideologies that view workers through the lens of the market and the goals of private enterprise. These ideologies are examined in detail in Chapter 5 as we critique standard economic theories.

Capitalist countries, although they share a belief in private enterprise, do have belief systems that differ in some significant ways. These differences are reflected in the unique practices and laws that guide social behaviors. From Russia, where Putin has dictatorial powers, to New Zealand, with its strong democracy, there are a wide range of political and social arrangements in countries where some form of capitalism dominates. All, though, have laws supporting private enterprise and the individual accumulation of massive amounts of wealth.

There are some other commonalities. Most reject the notion that "royal blood" is a rightful source of power which entitles a few aristocrats to set and enforce the rules that govern the country. On the other hand, most also accept the notion that wealth (whether inherited or not) bestows power. Wealth, in a way, can be thought of as the "royal blood" running through the veins of capitalism. Wealth confers social rank, with rights and privileges. Social rank, sometimes legally enforced, is also linked to other factors and individual characteristics. There are or have been differences in rank between citizen and undocumented worker; between a child and an adult; between a commoner and a royal; and so on.

Under capitalism, the major source of wealth, and power, is private enterprise; and the unequal rights of business owners versus those of employees have led to the huge inequalities in wealth today (see Wolff 2021). There are many problems related to these unequal rights that capitalist ideologies attempt to justify as a necessary cost of the benefits received from capitalism. And the population seems to accept the notion that what is good for the long-term interests of business is good for the economy. And what is good for the economy is good for the well-being of all.

But consider the implications of the fact that employers have rights that their employees do not have. Business owners have the right to control how business revenues are spent after financial obligations are met. They can determine how much employees are paid, what investments are made, whether a worker will be fired or hired, and they have the final word on all major business decisions. Government, beholden to the powerful, protects and supports private enterprise. The end result of this protection of business owners' control of the net income from their companies is that wealth concentrates in the hands of a few plutocrats, who then attain immense power. Antitrust regulation has not prevented a growing concentration of U.S. industries. Two or three firms account for most sales in several markets. Amazon, for example, sells most books bought in the U.S.[16] Health insurance, appliances, office supplies, meatpacking, and more are all heavily concentrated industries.

Laws protect the "right" of businesses to spend their money as they choose, as long as they pay taxes and do not engage in illegal activities. They are able to dictate terms of employment and income for millions of people, and can intervene in elections and policymaking through strategic expenditures. They have a great influence on foreign policy and military ventures. They alter the nature of human relationships and connections, particularly during times of crisis.

Globalization Impacts National Economies Differently

Capitalism, through transnational businesses in particular, drives today's global economic realities and so impacts the lives of millions. Financial and product flows, and flows of people, knowledge, and culture across the world, impact every nation. International corporates are prime movers of global developments—economic, social, and scientific. They shape international political relations. Flows of direct foreign investment between countries give an indication of the importance of economic ties between countries. Foreign investment flows totaled an estimated \$1.5 trillion in 2019, an amount equal to about 2% of the world's total GDP (\$87.3 trillion).[17] These annual flows, of course, lead to a much higher cumulative value of investments. The U.S., for example, has assets valued at \$6.2 trillion abroad as of 2020—an amount equal in magnitude to almost 30% of U.S. GDP (\$20.9 trillion in 2020).[18]

Large inflows of foreign capital investment have a significant impact on output and employment in recipient countries. In Mexico, a large percentage of total production is done by U.S.-owned or -controlled businesses.[19] Mexico received over \$110 billion in direct investments from the U.S. in 2018 alone, an amount almost equal to 10% of its total output of \$1222 billion.[20] These investments in industry and agriculture have consequences for the local economy that can be negative. U.S. agricultural investments may reduce local access to water and often decrease local ownership of land, with adverse effects on small farmers.

U.S. companies are currently holding an estimated $2.1 trillion in retained profits abroad. By retaining profits abroad, companies avoid the U.S. taxes that would be due upon repatriation.[21] After a change in tax laws in 2017 to encourage repatriation, over $1 trillion in taxes was collected on repatriated profits!

To understand why foreign investors are not always welcomed by recipient countries, it is important to recognize not only the inequality in power between the countries involved, but also to recall the often unsavory history of colonialism that still affects the present.

Fewer than a hundred years ago, several countries had colonies, and even today some small countries remain under foreign rule. The U.S., for example, still holds political control of Puerto Rico, Guam, the U.S. Virgin Islands, American Samoa, and the Northern Marianas Islands. Even after overt colonial ties have been severed, strong political and economic influence persists between former colonies and their former rulers. Former colonies often still depend on loans, purchases, and investments from the former ruler, and in many cases have little negotiating power to improve contracts and agreements.

Indigenous peoples in North America, Australia, and around the globe were among the victims of colonization; any reparations made do not come close to just settlements. With few exceptions, Indigenous peoples have had their land forcibly taken from them and still lack control of the resources that have been central to their cultures for centuries. Their sovereignty is limited, and threatened. At some times and in some places, they have been forbidden to use their native languages and their children have been taken from them to be educated into the language and culture of their conquerors. Their rights, even as stated in the laws of the countries where they live, are not always upheld, particularly when doing so might hurt the profitability of businesses.

The U.S. and other leading world powers continue to intervene in other countries, both directly and indirectly. The U.S. often intervenes if it finds that U.S. interests are at risk. This may be to prevent war in a region, for example, or to protect access to valuable resources. Democratically elected leaders have been overthrown by coups with clear U.S. involvement—the most well-known instances being the overthrow of Allende in Chile in 1973 and the 1953 overthrow of Mossadegh in Iran.

Both leaders had plans to nationalize leading industries—copper in Chile (U.S. companies affected) and oil in Iran (British interests affected).

Greece provides a recent example of how intervention by global economic powers can cause tremendous suffering in countries that depend on their assistance during times of crisis. The financial crisis of 2008 wreaked havoc on the Greek economy and the loan package it received as a bailout was conditioned on the acceptance of draconian austerity measures. These measures involved massive layoffs in the public sector including in health care, deep (over 20%) cuts in unemployment benefits and similar cuts in pensions, a lowering of both the minimum wage and private sector wages by 20%, tax hikes, and more.[22] These measures, rather than enabling the country to meet loan payments, threw the country into deeper recession and chaos—one of the worst recessions experienced by a country since World War II—and accordingly made further loans necessary. Reporters during the depth of the crisis painted a grim picture of widespread suffering and protests.[23] The International Monetary Fund, the main author of the austerity package, later acknowledged (2013) that the tax hikes and spending cuts it imposed were more than the economy could bear.[24]

Politically charged interventions into lower-income countries are not as common as other market interactions. Market forces at work between richer and poorer countries have mixed results. Jobs are created, but working conditions and pay associated with those jobs may be dismal or exploitative. We have already discussed the child slave labor on Nestle's cocoa plantations in Chapter 3. Worker protections in poorer countries are typically weaker than those in wealthy nations, and pay can be low even by local standards, particularly for women and, in some cases, children. In Bangladesh, fires in factories resurface as a problem every few years—a horrific one in 2013 killed over 1100 workers producing garments for Gap, Inc. and other clothing retailers. Exits were blocked.[25] Union leaders in poorer countries are often at risk: several union leaders in South America, Asia, and Africa have been killed, including those working in companies that are controlled by or subsidiaries of foreign firms such as Coca-Cola.[26]

The ideological messaging of the ruling elite in poorer countries, not surprisingly, stresses independence, autonomy, and greater local control of valuable resources. Even when these elite are authoritarian and repressive, their calls for greater independence resonate. Their messages describe the benefits of strengthening businesses owned by nationals, not foreigners, claiming that a wealthier national business sector is key to opening economic opportunities for all.

Radical Change: Past and Present

History makes clear that nothing stays the same. The U.S. power elite are always dealing with new technologies, new industries, and new billionaires. They respond to pandemics and terror attacks. Our plutocrats are continually adapting their messages to new situations—a task they accomplish with the help of the knowledge workers they fund. Their ability to shape the public's understanding of new developments through political, educational, and media networks works to their benefit and also serves to maintain social stability under changing circumstances. The hegemony of the ruling oligarchy, then, is ever-evolving.

There is little to no consideration of transformational change as a real possibility in the media, so it is not a surprise that the belief that "Real change is impossible" is widespread. This works to the advantage of those in power. At the same time, it is also true that major change brings upheaval and uncertain outcomes. This risk is the downside to meaningful reforms. Those who would lose power and those who would benefit both tend to resist major change. Changing the people at the top and the rules guiding behavior in organizations can be destabilizing and, at the least, puts an end to the security that comes from on-going routines. The downside to the status quo, of course, is that it ensures that those in power stay in power, and continue to pursue their priorities with the consequences we know only too well: widespread, unnecessary suffering, and tremendous environmental destruction. The major advances in social programs and other areas during the past century came only at great cost and should warn change agents to plan for battle.

Radical Change Historically: Europe in the 1900s

Up to the present, neither political parties nor technological change nor moral enlightenment nor the cooperative, socialist, and other movements have succeeded in ending the political dominance of the wealthy under capitalism. Much has changed in the past 150 years, however, and much has been learned along the way from people's efforts to create a better life for themselves and others.

The history of Europe since 1850 offers a treasure-trove of insights. The governments of European countries experienced dramatic changes particularly between 1900 and 1950. Each country followed a torturous political path. Some went from monarchy to strongly representative democracy via detours that at times included colonial empires, authoritarian rule, or dictatorship. Others began as democracies but gave way to dictatorship before righting course. In any case, the nature of the ruling class changed markedly over the period, impacted by two World Wars and the Great Depression; the number and kind of grassroots organizing efforts changed as well. In every country, labor fought political battles for greater rights and protections. Unions, benevolent societies, religious bodies, and others demanded that government address the needs of families and children. The advent of what is called "the welfare state" dates from this period. Unlike in the nineteenth century, it is now taken for granted that the government has a responsibility to protect and advance the welfare of citizens. The rise of the welfare state brought a vast expansion of social services that today in Europe typically includes universally accessible child care and health care, low or free tuition college education, paid family leave, and more.

Each country in Europe has a different history, but each experienced the industrial revolution and the accompanying changes in work-life. These changes brought worker resistance to low wages, poor working conditions, and to the effects of the new realities on family and social life. Across Europe there were increasingly organized efforts by working people to gain better living and work conditions. Political parties were created that took up these demands and unified these grassroots efforts.

Labor and socialist parties existed in all major countries by 1890. The demands of these parties for an end to the unjust exploitation of workers resonated not only with those in labor unions but with the broader population. Some of these parties eventually rose to power, although the base of these parties and the content of party platforms changed over time. Today many labor and socialist parties in Europe are mainstream. Still, progressive welfare reforms everywhere have their roots in workers' (including farmers') organized demands. At times, authoritarian rulers have instituted programs to the benefit of families and workers, but historians usually see these reforms as pre-emptive moves. For example, in 1883, Bismarck introduced a limited health-care scheme in Germany, in large part to reduce the appeal of socialist organizations. More commonly, progressive political parties were able to get legislation through national parliaments. Sweden, a poor rural country in 1900, was one of the first countries to become a welfare state: the Social Democrats in power in the 1930s and 1940s pushed for and got programs aimed at providing universal health care, a universal "people's pension," family allowances, and unemployment insurance.[27] The U.S. responded to widespread and militant labor unrest in the 1930s with the package of programs known as the New Deal. The Labour Party in the U.K. passed an ambitious slate of welfare programs after World War II. In other cases, dictators from Mussolini and Franco to Hitler instituted limited social programs addressing workers' needs in an effort to fend off the political threat posed by organized labor and socialist parties.

In the U.S. during the 1930s, unions grew in strength and millions of workers participated in strikes. The labor movement worked through the Democratic Party to gain the right of collective bargaining, a minimum wage, a maximum 44-hour workweek, the creation of social security, the prohibition of child labor, and other critical pieces of legislation. Unlike in Europe, though, labor did not unite behind the creation of a new political party. There was widespread interest in the prospect for a while, but union leadership was divided, and the legislative advances embodied in the New Deal under Franklin Roosevelt and Frances Perkins served to dissuade most labor leaders from further pursuit of the idea.[28]

Political developments since World War II have brought fluctuations in spending on hard-won social programs like healthcare and unemployment benefits as well as struggles over their content in both the U.S. and elsewhere.

Creating a Deeper, Radical Democracy Today

A key take-away from this overview of the emergence of Europe's progressive social programs is that these radical policy innovations were made possible because of grassroots pressures on political systems. During the twentieth century, most of Europe saw the emergence of deeper democracies that better served their citizens. In many countries, changes included government funding of social media that improved candidates' access to the voting public, as well as the institution of multi-party elections with proportional representation as the outcome—ensuring a wider diversity of voices in parliamentary bodies than is the case in two-party systems. (Chapter 6 discusses some possible political reforms for the U.S., based on European experiences.) Not surprisingly, the rise and fall of political parties and policies over time has meant fluctuations in the quality and quantity of social programs, but in Europe belief remains strong in universal access to health care, in the provision of assistance during spells of unemployment and ill health, and much more.

The history of Europe and the U.S. over the past 100 years, then, shows that major change in government programs is common, but that it comes at high cost.

Our political future depends on the response of national leaderships to the urgent needs of our time. These include the need for adequate economic security and opportunities; for an end to global warming; for an end to racism, misogyny, and other social oppressions; for the preservation of species; for needed care—health care, child care, elder care, and other; and for representative and accountable public officials. These projects are termed radical: they cannot be met without radical change in today's political structure with its focus on goals set by the powerful. Political theorists including Chantal Mouffe and Ernesto Laclau believe that radical projects can form the basis for coalitions working for a radical (deeper) democracy.[29]

As envisioned by these theorists, this transformed, radical democracy would be guided by beliefs and practices that are responsive to people's needs. Traditional democratic institutions would be strengthened, including multi-party politics, a free press, universal suffrage, trial by jury, and more. Radical democracy would prioritize meeting urgent demands. This deeper democracy cannot emerge from reforms that follow the usual institutional channels, but rather requires the mobilization of grassroots support for transformative change. Such grassroots coalitions would be united through a shared opposition to the rule of the plutocrats.

The laws, practices, and institutions of radical democracy would be rooted in the principles that motivate today's social movements. These include equality, liberty, justice, love of neighbor, and freedom of speech and debate (Mouffe 2018; Nussbaum 2004). These may be traditional values, but their concrete expression as interpreted by radical leadership would bring profound change: businesses would voice support for serving people over increasing profits, politicians would honor the popular voice over that of wealthy supporters, and so on. The policies implemented would bring solid progress toward ensuring that everyone has access to a satisfactory standard of living, that the planet's ecosystems are restored to health, and that other goals improving how we live together are achieved. There would be heated debates over policies, but the voices of all groups acting in good faith would be respected. Political agendas would not be set by the wealthy elite; debate would be agonistic, not antagonistic. Injustices suffered by those affected by multiple oppressions would be acknowledged and addressed.

Unlike today, public debates would be over alternative policies, any of which would bring positive change. Today, incremental improvement is typically what we hope for. Expectations would be higher under radical democracy. A unified, committed coalition among people seeking in-depth change is a pre-requisite to such a transformation of politics. This unity, of course, is difficult to achieve. Only by working together for shared goals over a long period of time can groups build the trust that is critical to merging persons with disparate visions into a unified force.

Change Begins with Challenging Prevailing Ideas

This process of transformative change begins with ideas that effectively counter prevailing notions. These new ideas can coalesce into shared ideologies that grow in popularity and unify efforts to achieve radical goals, including the establishment of a deeper/radical democracy. The content of ideologies that can undermine the current hegemony is hinted at by the challenges now being mounted to capitalist ideologies.

The belief in business and its pursuit of profit is perhaps the most fundamental belief sustaining the power of the elite. Profitable businesses, it is claimed, are the bedrock of a healthy economy and benefit employees, customers, and communities. If we are to envision an alternative economic reality that brings a shared prosperity, it is well to recognize the problems that originate with the rights and privileges bestowed on the wealthy by our laws and practices.

Consider that the wealth accumulated at the top is created by the many laboring below. Those who invented the products being sold, those whose scientific research was used to create those inventions, and those who produce and sell the products have no legally recognized rights to the income generated by their labor. Whistleblowers calling out harmful practices and union organizers have little effective protection against being fired under the law (Maltby 2009). Democracy at the workplace is very limited; many stakeholders in a business from employees to consumers have no say in its operations. On the other hand, business rights are protected, not only by law, but also by the reluctance of politicians, and much of the public, to put any additional restraints on business.

The result is that business rights are permitted to take precedence over human rights and the survival of the planet, with little or no penalty.

To state the obvious, the abuses of power by business are not effectively restrained by competition or regulation, and that includes antitrust regulation.[30] For a while after World War II, when unions were strong and an expanding economic pie benefited most, inequalities in wealth and income seemed manageable so that there was less incentive to change the laws that severely constrain the rights of stakeholders [i.e. employees, consumers, suppliers, communities as well as shareholders].

That is no longer the case. The rules of the "economic game" governing the marketplace have led to the results we have today. Winners win big and losers are left to fend for themselves. There are some different players—more women and Blacks—but the game hasn't changed. The relative power of those pushing back against a rigged system has dropped, with the consequences apparent. And decision-making power at the highest levels remains firmly in the hands of the wealthy.

Personal Political Beliefs Involve Both Reason and Passion

Changing one's beliefs about political parties or about the role of business bears some similarities to changing one's religious beliefs. Logic plays a role, but so does emotion. In fact, setting priorities always depends not only on facts and logic, but also on affect, even for simple tasks like drawing up a day's schedule. Since the 1990s, research on the functioning of the brain has made it clear that day-to-day functioning requires the jolt provided by our emotional reaction to information. Neuroscientists such as Damasio have shown that emotion is a critical element in sorting through the daily barrage of information. Emotions enable a person to set priorities in what would otherwise be an over-determined system. Brain-damaged patients, who can easily solve mathematical problems and other complex rational tasks, are incapable of allocating their use of time in a way that allows them to accomplish daily tasks in an appropriate manner and so have been unable to hold jobs that require that time allocation skill (Damasio 1994). Assigning varying degrees of importance to different tasks is done easily by persons with normal brain function, however. Changing political beliefs is a task involving the whole person. A person changing political beliefs must feel comfortable and accepted within his or her new social environment. In the case of breaking from capitalist ideology, changed attitudes toward the protections/rights of business owners must be accompanied by trust and belief that new forms of business can function well. This means a shared understanding with those working for such a new, more democratic economy.

Replacing Oppressive Ideologies

As we have seen, today's ideologies justify and protect the wealth and power of plutocrats. Challenging the plutocracy requires debunking the myths these ideologies perpetuate. Let's start doing that now.

Ideological arguments for capitalism, and its plutocrats, stress its economic benefits. These benefits are said to include a proven ability to expand output and raise average income levels. Private enterprise, it is also claimed, fosters the entrepreneurial spirit, individual initiative and innovation, and so raises living standards. These benefits, it is feared, would be lost if the rights of business were curtailed and businesses subjected to greater democratic control. These arguments have considerable traction. Responding to them requires insight into how the economy works, but also into how people perceive today's economic realities.

We begin by stating our goal, namely, to reshape society into a vibrant and egalitarian democracy. This requires reshaping the economy. Major economic decisions that affect the lives and livelihoods of all should not be left to the plutocrats and their minions, as is now the case. But is it possible to increase democratic say in the economy while at the same time ensuring that living standards improve and that the drive to innovate and create is fostered? What would be the likely effects on output and incomes of reducing rights now claimed by owners of big business? What would greater political power for the average citizen mean for the economy? Understanding how to reconfigure the economy successfully requires understanding the basic strengths and limitations of today's capitalism. Where are tweaks needed? Where must new organizational structures be put in place? And how can people be won over to the idea of greater economic democracy? We tackle these questions in the remainder of the book.

Technological progress that improves living standards and restores the environment is one key ingredient of a better economy. An important and generally accepted conclusion reached by economists is that the high levels of productivity (output per worker) and of incomes today are due largely to technological change, not to increased investment in existing technologies. Innovation, then, should be fostered. Innovative companies can be private, for-profit enterprises or they can be organized under alternative ownership structures, such as cooperatives or ownership by a trust. [The pharmaceutical giant Novo Nordisk is owned by a trust as is the conglomerate owning Jaguar. Trusts that own large companies were typically set up by the business' founders or owners to provide oversight of operations after they stepped down from active leadership roles.]

The conclusion is that the fostering of research and innovation is basic to progress in living standards and to ecological health, and should be a focus of any economic reforms undertaken by a more democratic government.

There are other considerations. By examining what economists have to say about the market in some detail, we can better understand how to build on its strengths and counter its weaknesses. In Chapter 5, we focus on basic economic theory regarding shortcomings of the market and the appropriate role for the government in the economy. We also look at some critiques of standard approaches from environmentalists, feminists, and others. We question the neglect in economic theory of power, which plays a critical role in the economy, and briefly present the argument for recognition of stakeholder rights in businesses [stakeholders include employees, suppliers, and the community as well as shareholders]. Chapter 6 examines some successful policies in place around the world.

The Market Economy: Pluses and Minuses

A key message of the ideology supporting capitalism for over two hundred years is, in brief, that market forces guide an economy to prosperity. *The Wealth of Nations*, written by Adam Smith in 1776, describes how the invisible hand of the market efficiently balances the supply and demand for goods in a way that benefits both consumers and producers. This market-led system of production, Smith argued, is the source of the wealth of nations.

Even that renowned anti-capitalist, Karl Marx, recognized the tremendous economic achievements of capitalism, as seen in this excerpt from *The Communist Manifesto* by Marx and Engels (1848):

> It (capitalism) has accomplished wonders far surpassing Egyptian pyramids, Roman aqueducts, and Gothic cathedrals; it has conducted expeditions that overshadow all former exoduses of nations and crusades.

But the destructive power of capitalism is also indisputable. Periodic recessions and depressions shutter thousands of businesses, strangle production and consumption, wreaking widespread suffering. The low prices prevailing for many goods benefit consumers, but the associated low wages hurt workers, particularly workers drawn from marginalized populations. The true value of air, water, ecosystems, caring labor, and much more are not recognized by the market. Economic incentives to protect most natural resources, flora and fauna are weak or nonexistent. Pillaging of the earth and an undervaluation of human bonds have been the result. Ecosystems have been deeply degraded across the globe; thousands of species have become extinct or are close to extinction. Global warming is bringing sea-level rise, changing weather patterns, and threatening the future of humankind.

A Closer Look at Organized Struggles Against Capitalism
Efforts to end the abuses of power under industrial capitalism date back to the latter half of the 1700s when unions and workingmen's associations were formed. These worked for better pay and working conditions and to meet other needs of their members. The ideas of socialism, cooperatives, communes, and other alternatives to capitalism were introduced and spread across countries. By the mid-nineteenth century there were hundreds of cooperatives and communes, found in many countries and continents.[31] The economist Alfred Marshall (1842–1924), the creator of supply and demand curves and a founder of neoclassical economics, was a supporter of cooperatives, even serving as president of the U.K. Co-operative Congress in 1889. More important to political agendas, though, were socialist and labor parties, which persist today throughout the world in some form.

Unlike in Europe, the labor movement in the U.S. did not unite behind the creation of a political party. There was widespread interest in the prospect, but the Democratic party sought to head off the formation of a labor party and offered various concessions to workers, many embodied in New Deal legislation associated with Roosevelt. As part of the New Deal, the Department of Labor under Frances Perkins undertook several initiatives that improved the outlook for workers during the Depression years.

Socialist, Communist, and Labor parties have been leaders in the resistance to unbridled capitalism. Karl Marx was the guiding theoretician of the left wing of the socialist movement and the leading voice in the First International, also known as the International Working Men's Association (1864–1876). His 1848 *Communist Manifesto*, written with Friedrich Engels, held that workers' revolutions could bring about a workers' state in capitalist countries. Meetings of the International were convened to develop strategies to achieve this goal. Communist and socialist parties were formed around the world with this same goal. Marx died in 1883, before the Second International (1889–1914), which included representatives of all the socialist and social democratic parties and trade unions in Europe, the U.S., Canada, and Japan; over 20 countries were represented, with almost 9 million voting members.[32] That Congress eventually adopted the position that achievement of socialism through gradual reforms was not possible and rejected cooperation with nonsocialist or non-labor parties, although it retained belief in parliamentary democracy. Division over these and other issues split the Congress and led to its end. Successor Congresses, dominated by Moscow, changed the guiding principles and goals of the Congress.

Concluding Remarks

Our democracy has been hijacked by plutocrats whose influence has an undue impact on our nation. These plutocrats dominate our industries and institutions. Their power is shored up and enhanced by ideological messages that are continually updated and dispersed through social media. They are often ruthless in their pursuit of wealth; many go to great lengths to oppose unionization, increased worker protections and pay, or environmental regulations. They strongly influence legislation, elections, and social values. The prevailing pro-business ideology supports their right to exercise power and even gives them sufficient clout to win government interventions in other countries, at times military intervention, should those countries severely restrict their business operations.

The first step toward replacing this plutocracy with a deeper (radical) democracy is to recognize their hold and counter the ideologies that protect them—offering an alternative vision of a caring, cooperative, and prosperous society.

The most fundamental value upholding capitalism and the capitalist plutocracy is the belief that holds sacred the right of capitalists to control the wealth created by the enterprises they own. There is a huge gap between the ideals of a capitalist economy, which professes to offer prosperity to all willing to work and its actual functioning. Similarly, the ideal of a democracy of, by, and for the people is far from being realized; in fact, the people have little ability to ensure that businesses behave responsibly or that the politicians they elect represent their interests.

There are several ideologies that justify the current ways of functioning. Each institutional system, from religions to the military, from the family to the presidency, has ideals and beliefs that guide behavior. These ideologies may offer idealistic beliefs about freedom and fairness, while at the same time espousing oppressive beliefs about different groups of people or about animals and the natural world. Oppressive beliefs have negative consequences, both for the victims of those beliefs and for society as a whole. Oppressive beliefs create division and discord, which are formidable obstacles to creating the trust and common cause needed for a society to flourish. Divisions between Whites and people of color, between men and women, between gays and straights, and so on have long made political organizing for positive change more difficult. Established powers benefit as a result. During the past 100-plus years, there has always been push-back against the abuse and misuse of power by business. Laws against child labor, against unnecessarily dangerous work conditions, and more have eventually made it through Congress and been upheld by the courts. Still, the power of the plutocrats remains excessive. The few at the top of a company continue to have the right to determine the distribution of income created by the many, without the consent or even input from the co-creators of that income. Business, as part of our democracy, should be held accountable for its behavior. Business operations must evidence a stronger commitment to sustainability and to human rights. Many business professors recognize this and are re-visioning the corporation in light of stakeholder rights that have too long been neglected or denied. Stakeholders, including the broader community, deserve a say in how business is conducted. Greater corporate accountability is imperative to moving us forward.

If our democracy is to be made deeper—more participatory, more accountable, and more transparent—the ideologies that justify the domination of plutocrats and the oppressive divisions among people must be confronted. Battles today for deeper democracy are reminiscent of the battles of labor and of others in years past. Change is possible. Across time and across countries, we have seen that shifts in power away from the elite occur, but only with widespread engagement in political struggles. In the next chapter we look more deeply at how ideologies distort the true nature of the power realities shaping life in the U.S. today and at some of the alternatives we face.

Notes

1. https://www.cbsnews.com/news/poll-americans-say-money-has-too-much-influence-in-campaigns/.
2. https://www.pewresearch.org/fact-tank/2020/07/22/most-americans-say-social-media-companies-have-too-much-power-influence-in-politics/.
3. https://news.gallup.com/poll/193484/americans-major-donors-sway-congress-constituents.aspx, https://news.gallup.com/poll/188747/majority-americans-dissatisfied-corporate-influence.aspx.
4. Hegemony is a concept that is associated with Antonio Gramsci (2011) who developed current understandings of the term while writing in prison (1929–1935) as he grappled with the question of why the idea of a workers' state had failed to fire the commitment needed for successful action in Europe. https://www.newstatesman.com/culture/2018/02/why-antonio-gramsci-marxist-thinker-our-times, lays out some of the widely recognized intellectual contributions of Gramsci. Other work on power builds on Gramsci's insights: see Foucault (1989), Fraser (1989), Haslanger (2012a, 2012b), Laclau and Mouffe (2014), and Hill (2003).
5. The shared knowledge that influences society is changed on a day-to-day basis by those in authority. For example, the initial onset of the COVID-19 pandemic was met by rules regarding isolation and social distancing. Those not in authority have much less ability to influence policies and laws. Large protests targeting a specific demand usually will be met with a placating response with little substance; until recently, this was the case in instances of police killing persons in custody, often Black men. The definition of poverty is, in effect, three times the cost of adequate food for a family; the definition of inflation used by the Federal Reserve is based on a bundle of goods and services that does not include the cost of rent. If these definitions were modified to more accurately reflect families' economic situations, the poverty rate would be higher and the soaring cost of rent would impact the Fed's decisions. Political considerations are undoubtedly one reason why these definitions have not been changed.
6. See Kreitler https://www.researchgate.net/publication/292882625_.
7. Many of us depend for our livelihoods on jobs that result in more profits and greater wealth for business owners and plutocrats. Even small businesses often rely on contracts with corporations and other purchases by big business. They may hold a franchise. (Most small businesses rely on big business.) Those of us striving to move up the career ladder or improve our earnings will attempt to act, and think, in ways that will bring success. We integrate ourselves into the group and become personally invested in achieving the common goals.

   Others of us are in low-income jobs taking care of the dependents of others, while we ourselves may lack health care coverage and rely on food stamps to survive. Some of us serve in the military, where we risk giving our lives in battles too often waged to protect U.S. interests. We purchase products from corporations that mistreat and underpay employees and subcontractors, and that cause environmental degradation.

Some of us in corporations are using our creativity and intelligence to improve the products that add to the bottom line of our firms; others of us are working in government agencies enforcing legislation that is meant to protect people and the environment. We may work with nonprofits or firms that treat employees and customers very well. Whatever our job, we usually recognize the worth of what we are doing, even if we may want to change our jobs in some way.

8. https://aeon.co/essays/how-the-frankfurt-school-diagnosed-the-ills-of-western-civilisation, explaining away everything via reason has been destructive of spirit.
9. https://www.theatlantic.com/sexes/archive/2013/03/the-difference-between-a-happy-marriage-and-miserable-one-chores/273615/. The fair sharing of less pleasant jobs makes for better relationships at home, and at work.
10. https://www.epi.org/publication/black-white-wage-gaps-expand-with-rising-wage-inequality/.

    Key findings include that there is a pay gap between white men and men of color with similar backgrounds: "relative to the average hourly wages of white men with the same education, experience, metro status, and region of residence, Black men (as of 2015) make 22% less, and Black women make 34% less". Much of this pay gap is likely explained by discrimination in hiring, pay increases and promotions, although such discrimination is difficult to quantify. In one of the experimental studies corroborating the common perception of discrimination, researchers changed the names on resumes and analyzed the results in terms of callbacks for interviews. There was a significantly higher callback rate for Emily and Greg than for Lakisha and Jamal: https://www.nber.org/papers/w9873. On the pay gap for men with a bachelor's or more: https://www.payscale.com/research-and-insights/racial-wage-gap-for-men/.

    George M. Frederickson (2002) pointed out that belief in equal rights while denying those rights to some requires some justification—typically that those denied rights are inherently deficient—different from the norm. This is related to the recurring although thoroughly debunked belief that people are poor usually because of some personal deficiency or defect of character.
11. https://iwpr.org/media/press-releases/women-are-paid-less-than-men-in-almost-all-occupations-according-to-new-analysis-of-2020-earnings-data/.

12. https://www.levyinstitute.org/pubs/pn_12_01.pdf on Greece.
13. https://bjs.ojp.gov/content/pub/pdf/fvv.pdf.
14. https://williamsinstitute.law.ucla.edu/press/ncvs-lgbt-violence-press-release/. "LGBT people are nearly four times more likely than non-LGBT people to experience violent victimization, including rape, sexual assault, and aggravated or simple assault, according to a new study by the Williams Institute at UCLA School of Law."
15. Maltby (2009: 256). See also www.workrights.org: "Can your boss fire you for a bumper sticker on your car? smoking or drinking in your own home? what you write on your personal blog? refusing to answer questions about your sex life? If you answered 'no', prepare to be shocked."
16. https://www.wbur.org/radio/programs/onpoint/tag/monopolies-in-america, https://www.openmarketsinstitute.org/learn/monopoly-by-the-numbers, gives data on the concentration of market power in several industries e.g. Walmart, Amazon dominate several markets.
17. https://unctad.org/news/global-foreign-direct-investment-fell-42-2020-outlook-remains-weak.
18. https://www.mayberryinv.com/u-s-direct-investment-abroad-rose-to-6-15-trillion-in-2020/, citing data from the U.S. Bureau of Economic Analysis. U.S. capital stock was $69 trillion in 2019. https://fred.stlouisfed.org/series/RKNANPUSA666NRUG.
19. The OECD (Organization for Economic Cooperation and Development) criteria for determining whether a financial transaction qualifies as direct foreign investment is that the investment bring substantial control of a foreign-based business—a rule of thumb for statistical purposes would be a minimum 10% ownership stake in that company. https://www.investopedia.com/terms/f/fdi.asp. For more detail, see https://www.oecd.org/daf/inv/investment-policy/fdibenchmarkdefinition.htm.
20. https://www.state.gov/reports/2020-investment-climate-statements/mexico/. The United States is Mexico's top foreign investor, investing $115 billion in the country in 2018 or 40% of all inflows to Mexico. Mexico's GDP in 2018 was $1222 billion. U.S. investment over the years, then, accounts for a large percentage of Mexico's output. U.S. companies are major players in many countries around the globe.
21. https://groco.com/article/how-much-profits-are-u-s-companies-really-holding-overseas/, https://www.latimes.com/business/story/2019-12-19/companies-repatriate-1-trillion-since-tax-overhaul. Cites Bloomberg as source.
22. Polychroniou, C. J. "Neo-Hoovrian Policies Threaten to Turn Europe into an Economic Wasteland". Levy Economics Institute of Bard College, *Policy Note*, No. 1, 2012.

23. www.theguardian.com/world/2015/jul/09/greek-debt-crisis-damage-healthcare-hospital-austerity, www.nytimes.com/2011/10/20/world/europe/greek-workers-start-two-day-anti-austerity-strike.html.
24. Papadimus, Lefteris, and Renee Malterzou. 2013. "For Hard-Hit Greeks, IMF Mea Culpa Comes Too Late". *Reuters, Business News*, June 6, 2013.
25. https://www.theguardian.com/sustainable-business/rana-plaza-gap-worker-safety.
26. https://www.pbs.org/frontlineworld/fellows/colombia0106/video_chapter1.html, https://www.solidaritycenter.org/tag/murdered-trade-union-leaders/.
27. https://www.rug.nl/ggdc/historicaldevelopment/maddison/releases/maddison-project-database-2020, gives historical data going back before 1900 that shows the turn-around in countries like Norway, Finland and Sweden. Misgeld et al. (1992) provide more background.
28. Preis (1974: 47, 475).
29. https://literariness.org/2018/08/14/the-political-theory-of-ernesto-laclau-and-chantal-mouffe/. Nasrullah Mambrol offers a far-ranging, insightful discussion of political theory, touching on Foucault, Althusser, Derrida, Lukacs, Heidegger, Badiou and more in relation to Laclau and Mouffe's poststructuralist understanding of the political.
30. https://www.wbur.org/radio/programs/onpoint/tag/monopolies-in-america, discusses antitrust efforts, past and present.
31. Fourier is the best-known leader of the movement to create communes.
32. Entry for "Second International", Britannica.com.

References

Damasio, Antonio. 1994. *Descartes' Error*. London: Penguin.

Economic Intelligence Unit. 2006–2021. *Democracy Index* (annual). London. https://www.eiu.com/n/campaigns/democracy-index-2020/.

Foucault, Michel. 1989. *The Archaeology of Knowledge*. London: Routledge.

Fraser, Nancy. 1989. *Unruly Practices*. Minneapolis: University of Minnesota Press.

Frederickson, George. 2002. *Racism: A Short History*. Princeton, NJ: Princeton University Press.

Freeland, Chrystia. 2013. *Plutocrats*. New York: Penguin Books.

Gilens, Martin, and B. Page. 2014. Testing Theories of American Politics: Elites, Interest Groups, and Average Citizens. Published online by Cambridge University Press, September 18. https://www.cambridge.org/core/journals/perspectives-on-politics/article/testing-theories-of-american-politics-elites-interest-groups-and-average-citizens/62327F513959D0A304D4893B382B992B.

———. 2014. Testing Theories of American Politics: Elites, Interest Groups, and Average Citizens. *Perspectives on Politics* 12 (3): 564–581. https://doi.org/10.1017/S1537592714001595.
Gramsci, Antonio. 2011. *Prison Notebooks (Vols. 1, 2 & 3)*. New York: Columbia University Press.
Haslanger, Sally. 2012a. *Resisting Reality: Social Construction and Social Critique*. Oxford: Oxford University Press.
———. 2012b. *Ideology of Sexism*, 1st ed. Oxford: Oxford University Press.
Hill, Marianne. 2003. Development as Empowerment. *Feminist Economics* (Summer): Special Issue on work of Amartya Sen.
Laclau, Ernesto, and Chantal Mouffe. 2014. *Hegemony and Socialist Strategy*, 2nd ed. London: Verso.
Maltby, Lewis. 2009. *Can They Do That?* New York: Penguin.
Marx, Karl, and F. Engels. 1848. *The Communist Manifesto*. https://www.marxists.org/archive/marx/works/1848/communist-manifesto/.
Misgeld, K., K. Molin, and K. Amark (eds.). 1992. *Creating Social Democracy*. University Park, PA: Pennsylvania State University Press.
Mouffe, Chantal. 2018. *For a Left Populism*. London: Verso.
Nussbaum, Martha C. 2004. Capabilities as Fundamental Entitlements. In *Capabilities, Freedom and Equality*, ed. B. Agarwal et al. New Delhi: Oxford University Press.
Page, Benjamin, Larry Bartels, and Jason Seawright. 2013. Democracy and the Policy Preferences of Wealthy Americans, January. A Russell Sage Foundation Report. https://www.russellsage.org/research/reports/wealthy-americans-political-preferences.
Preis, Art. 1974. *Labor's Giant Step*. New York: Pathfinder Press.
Przeworski, Adam. 2010. *Democracy and the Limits of Self-Government*. New York: Cambridge University Press.
Wolff, Edward. 2021. Household Wealth Trends in the United States, 1962 to 2019. Working Paper 28383. Cambridge, MA: National Bureau of Economic Research. http://www.nber.org/papers/w28383.

CHAPTER 5

Power, the Pursuit of Profit and Economic Theory

Economic theories shape many of the policies intended to solve national problems and achieve national goals, from healthcare coverage for all to net zero greenhouse gas emissions. Unfortunately, mainstream economic theories are not up to the job. The overview of economics given below, though brief, makes clear why it falls short: it embodies a narrow vision that limits and severely biases its findings and prescriptions. There is wide agreement among economists that changes are needed. I begin with a look at the microeconomic theory of markets as taught in college, and debunk some widespread myths about capitalism and the market economy. I then present a quick rundown of some of the exciting developments that are pointing the way to a new economics, one better suited to tackling today's challenges.

Economists on Market Economies

Efficiency that benefits consumers is a basic ideal of the market economy. As Warren Buffett puts it: "So we've got something that works in the market system…(it) turns out lots of goods and services people want."[1]

If you have ever taken an economics course, you have heard of the "invisible hand" of the market, introduced by Adam Smith in 1776 in his *Wealth of Nations*. This book presents the first formulation of market efficiency. Smith notes that an "invisible hand" coordinates the decisions of

M. T. Hill, *Building a Solidarity Society*,
https://doi.org/10.1007/978-3-031-07349-6_5

hundreds of consumers with those of individual producers. As producers pursue their own economic self-interest, each seeks to achieve the greatest output of their product at the lowest cost. At the same time, competition from other producers drives down the price of the product, which eliminates any excess profit and benefits consumers. The end result is that society as a whole obtains the maximum possible output of marketable goods and services at a price that is just above cost. An efficient outcome.

The field of *microeconomics* builds on the foundation laid by Smith.[2] It focuses on the operation of markets, laying out in mathematical splendor requirements that must be met to achieve efficiency. Neoclassical economists were the first to develop mathematical models that are now so basic to the field. Among their ranks was Alfred Marshall who, in his 1890 book *Principles of Economics*, used graphs to illustrate the workings of supply and demand. The accompanying figure presents the basic graph still used in teaching students how market competition leads to an efficient outcome. It shows a downward sloping demand curve QD, indicating that as the price of a good drops, the quantity demanded by consumers increases. The supply curve QS, on the other hand, slopes upward because suppliers will supply a greater quantity of a good as its price rises. Where the price consumers are willing to pay equals the price that producers are willing to accept, equilibrium is achieved: the quantity demanded equals the quantity supplied at the equilibrium price. This is an efficient outcome if certain conditions are met: markets are perfectly competitive, consumers are fully informed, consumer preferences do not change over time and are independent of the preferences of others, there are no transaction costs, and more. Changes in any of these assumptions can change the mathematics involved; and much of microeconomic theory today addresses such complications.

Notice that in Fig. 5.1 when the price of cheese rises above \$3 per ounce, the quantity supplied exceeds the quantity demanded. Similarly, if the price falls below \$3 per ounce, the quantity demanded is greater than the quantity supplied. Under the assumptions behind the graph, price affects both the quantity supplied and the quantity demanded. This reasoning has been used by those opposing subsidies, for example, to support the price of milk. According to their arguments, the subsidized prices will mean too much milk is supplied, diverting farmers from producing other goods. Similarly, restrictions limiting price increases, as in rent control, are argued to lead to shortages.

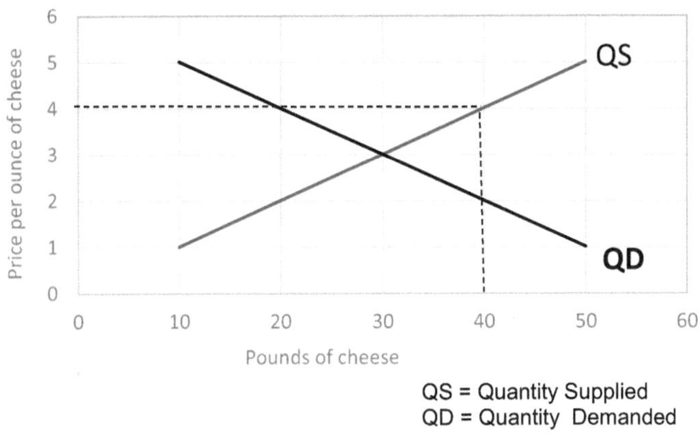

Fig. 5.1 Supply and demand for cheese

Demand QD and supply QS curves. The vertical axis of the graph gives the price per ounce of cheese, from \$0 to \$6 per ounce. The horizontal axis shows pounds of cheese, from 0 to 60. The line QD slopes down from the upper left toward the bottom right. The line QS slopes upwards from bottom left toward upper right. These two lines intersect at the point showing a price of \$3 and a quantity of 30 pounds. This is the equilibrium. There is also a dotted line indicating that when the price of cheese is \$4 per ounce, the quantity supplied will be 40 pounds, as shown by the line QS. But the quantity demanded at \$4 will be 20, leaving a gap between supply of 40 and demand of 20.

This simple story has been greatly elaborated since 1890. But the basic message of Smith remains at the heart of microeconomics: as long as perfect competition prevails, markets are efficient, benefiting both buyers and sellers. Efficiency rests on the low-cost production achieved due to competitive pressure as producers deliver goods to consumers based on their spending choices. This simple message has appeal. After all, markets do provide goods and services that consumers want if they can pay the price.

What About Worker Satisfaction?

You may have noticed that we have discussed market efficiency in terms of prices, quantities, production costs, and consumer satisfaction, but that we have not mentioned worker satisfaction. Neoclassical theory, in fact, treats the huge chunk of life spent on the job as merely a means of obtaining the income that enables the satisfaction coming from consumption.

Competition among workers in the labor market differs substantially from competition among producers, but this difference is not recognized in basic neoclassical theory. The institutions that protect the power of business owners and restrain workers' rights while on the job are taken as givens, assumed but not analyzed. Later in this chapter, we consider power differentials at the workplace and their consequences—something woefully neglected in mainstream economics. While the study of power is necessarily multidisciplinary (see Rorty 1992), the failure to include power in economic analyses reduces both accuracy and usefulness.

The Impossible: Perfect Competition

The markets where goods and services are bought and sold are postulated to be "perfectly competitive" in basic neoclassical economic theory. But what is perfect competition? Under perfect competition, a firm does not produce unique products—rather, an unlimited number of other firms can produce the same product. Corn, bread, eggs, and other products are examples of goods produced by many firms. Profits are a short-term phenomenon under perfect competition: there are no barriers to entry so, if a firm is profitable, other firms will enter the market, expand production, and drive down the price until only the cost of production is covered, thus eliminating any profit. (No barriers to entry means that everyone has equal access to the resources needed to start a firm.)

On the consumer side, it is assumed that each consumer spends his or her income in a way that maximizes his or her satisfaction or utility. As the consumer purchases more of a good, the satisfaction derived from each unit declines and consumers adjust their purchases accordingly. The basket of goods purchased will be such that the utility derived from the last unit of a good is similar for all goods purchased. There are other assumptions built into the model as well: a consumer's preferences don't depend on what other people are doing and don't change over time.

The equilibrium price that equalizes quantity demanded with quantity supplied can then be shown mathematically to minimize the cost of production and to maximize consumer utility. This outcome is termed efficient. Again, this result rests on the many assumptions built into the model. Perfect competition prevails; and both production cost curves and consumer utility curves have nice mathematical properties like continuity. Under perfect competition, financial markets are competitive as well, so that anyone with the necessary skills will be able to borrow money to start up a business and compete.

When perfect competition prevails, the inner workings of a firm are irrelevant: management decisions don't affect market outcomes because competition will weed out firms with inefficient practices. Firms could be cooperatives or corporations. Owners and managers may have a desire for personal power and wealth, but there is little they can do under the circumstances assumed.

Accordingly, no attention is paid to the relationships that exist within firms between employer and employees or relationships within households. The inequality in power of employer versus employee (or, say, husband versus wife) is not considered relevant.

Meanwhile in the real world… The world of perfect competition, of course, is not the one in which we live. Seldom do profits last only until other firms begin producing the same product. Often the market for a product is protected by barriers to entry, advertising, location, and patents or copyrights which provide at least a short-term monopoly.

The possibility that firms can attract consumers through advertising, that a firm's location can provide advantages, or that other factors can reduce competition and increase long-term profitability is addressed by theories of imperfect competition. These theories make it clear that markets are far from perfectly competitive and can be far from efficient. However, the policies proposed by economists to remedy such situations have had very limited success.

Economists, in any case, have identified ways to make markets more efficient. But more is needed from our economy than efficiency in production. Ask yourself: can the market ensure that everyone has access to a living wage? to health care? Can the private sector provide national defense? protect water supplies? Mainstream economics offers its answers to these questions and more and suggests some policies that can improve outcomes.

What Markets Can't Do

Economists acknowledge that markets can't do it all.

1. Markets cannot provide public goods such as national defense, super-highways, or clean air.
2. Nor do markets ensure that there are no costly negative effects on the environment or on people as by-products of market activities (These negative effects are termed negative externalities; there are also positive externalities, such as those that arise when people are vaccinated).
3. Markets do not prevent poverty or extreme income inequality.

These areas of market failure—public goods, externalities, and income distribution—have received a good deal of attention from economists. All agree that government has a role in addressing such problems.

The prevailing wisdom of economists regarding the role of government can be put succinctly: government should intervene in the economy only as needed when the market falls short. There are several tools at the government's disposal to improve market outcomes.

A Minimal Role for Government

When markets fail to deliver a good at a quantity and price deemed satisfactory, intervention may be needed. The government, in such cases, can either improve the functioning of the market (through antitrust legislation, for example) or can ensure that the desired good is provided, whether by providing subsidies, imposing taxes, instituting regulations or directly supplying the good. Tax breaks are a common form of subsidy. Governments provide a wide range of tax breaks to firms to encourage investment and job expansion. They also, in some cases, give direct subsidies to firms and individuals. Governments directly supply national defense, justice systems, legislation, enforcement of laws, and much more.

Governments impose fines to reduce pollution; they require that drivers pass a driver's test before receiving a license; and they take many other steps to reduce the negative effects of various activities. There is often disagreement among economists as to the best approach to tackling a problem. Consider the goal of reducing poverty or improving the income distribution. Minimum wage legislation, unemployment insurance, and the social security program are among the more accepted government interventions that have the effect of lifting incomes, but there is significant debate when it comes to other policy approaches.

How much government is too much? As we have seen, it is generally agreed that government should provide such public goods as national defense, contract enforcement, law and order, and education. But how much should be spent on the military? on parks? on education? To what extent should the government impose regulations to ensure consumer and worker safety? The answers to such questions vary by economist. The appendix to this chapter uses standard economic theories to analyze the market failures associated with Hurricane Katrina and the financial crisis. The analysis illustrates how it is that the apparently reasonable approach of economists to these crises neglects the elephant in the room—power disparities. But in fact power disparities play a large role in determining the economic fates of different population groups.

Billionaires on the Role of Government

Warren Buffett, like many of the wealthy elite, believe that economist Milton Friedman was right when he stated that "the social responsibility of business is to increase its profits." Business firms' first responsibility is to their shareholders, he argues. It is the role of government to pass and enforce legislation that holds business to society's shared moral standards. It is through government that we address society's shortcomings and failures. "The government has to play the part of modifying a market system."[3]

Other billionaires seem to share this view. There were 2755 billionaires as of 2021, worth $13.1 trillion in total, according to Forbes.[4] Here's what some of them had to say about how to address the problem of the huge and growing inequality of wealth in interviews with CNBC.[5]

Marc Benioff, an internet entrepreneur and CEO of Salesforce, calls for a "new capitalism," which among other things would include higher taxes on the richest, including himself. "Nationally, increasing taxes on high-income individuals like myself would help generate the trillions of dollars that we desperately need to improve education and health care and fight climate change."

Another billionaire, Mark Cuban, said "We as entrepreneurs have got to make a point to give stock to everybody that works for us. Period. End of story. No exceptions, because that's the only way people are going to get any type of equity appreciation." He continued, "Our country is a lot like running a business ... you can't just look at the short term in the immediate aspect, you've got to look at the long term."

The CEO of JPMorgan Chase, Jamie Dimon, lauded capitalism as the most successful economic system ever, but added that the governments of capitalist economies ought to support those who are struggling, recognizing, and rewarding work and providing safety nets for those unable to work.

If the suggestions above were followed, the income levels of many middle- and lower-income families would increase. These, and other suggestions, are not new, but the ideas have made little headway. The wealthy as a group seldom agree to such policies, claiming they would restrict their options and cut into their incomes. A few policies in this vein have been passed but with only modest impact, and inequality continues to increase.

The Rise of Macroeconomics, Synthesis with Microeconomics

When the Great Depression of the 1930s hit, neoclassical theory lost standing: it could not explain the prolonged recession. According to the original neoclassical paradigm, markets are self-correcting. Under this paradigm, if there is a recession and workers are unemployed, wages will fall until the hours of labor supplied match the hours of labor demanded. Only voluntary unemployment then remains—work can be had if workers are willing to accept the lower pay offered by employers. This analysis of the mechanics of recession was widely criticized. Wages that are below subsistence level mean that workers might not survive to make it to work.

Macroeconomics emerged as a separate field from microeconomics as a more compelling analysis of recessions gained acceptance. John Maynard Keynes in his 1936 book explained why recessions occurred and why market mechanisms failed to end them. He argued that the market clearing mechanisms that work at the micro-level of individual markets do not necessarily work at the macro-level of the economy as a whole. His influential theory led to what has been termed the neoclassical synthesis: microeconomic theories apply to the market behavior of individuals, firms, and households, while analysis of aggregate economic phenomena such as business cycles, inflation, and unemployment requires the tools of macroeconomic theories. Keynes' theory rests on characteristics of labor markets and financial markets: he argued that the interaction of demand and supply in those markets does not necessarily tend toward the full employment equilibrium. For one thing, workers require a minimum wage to survive which sets a limit on how far wages can drop in the economy as a whole; and interest rates similarly will seldom fall below zero.

By the 1970s, a school of thought termed "neoliberalism" gained prominence. The term is often associated with economist Milton Friedman and the Chicago School. Neoliberalism advocates market-oriented reform policies such as deregulation of markets, elimination of subsidies and of price controls such as the minimum wage, a lowering of trade barriers, privatization of government-owned assets such as the post office, and austerity policies for countries in debt.

Economists, whether neoclassical, Keynesian, or neoliberal, all rely heavily on mathematical models in their analyses, and use a wide range of mathematical tools and assumptions. Most variables in the models of market behavior are related to supply and demand, beginning with standard variables impacting price, such as cost, consumer preferences, resource constraints, and income distribution, but going on to include measures of other variables such as consumer expectations and irrational, yet predictable, behaviors. Descriptions of the work by Nobel prize winners in economics provide some idea of the wide variety of models now used in the profession. See, for example, the work of Tversky and Kahneman (1986) on decision-making, that of Simon (1991) on organizations, or that of Cohen (2005) on emotion and cognition.

Empirical verification of these models is de rigueur. This demands that variables used in the models be quantified. This often presents a problem: many variables of critical importance are difficult to quantify. What is the value of clean air or clean water? Of a healthy ecosystem? Of a human life? Of fulfilling, egalitarian relationships? How can we assess interdependent human relationships to determine their impact on: health, fulfillment, creativity, and incomes? How could a more egalitarian democracy be modeled and compared with our current democracy? Even with increasingly sophisticated models, economists struggle to grapple with such questions.

Economics, of course, continues to evolve. Economists have broadened the number of indicators they use to assess economies. No longer is per capita gross domestic product (GDP) the only economic measure found in country rankings. Life expectancy, income distribution, the average number of years of education, and even variables such as emissions of greenhouse gases may be provided in reports and studies. For a brief overview of the 17 U.N. Sustainable Development Goals (SDG) with 100-plus associated statistical measures, see Stiglitz (2020). Stiglitz also discusses a shortened flexible "dashboard" version of the U.N. SDG that he and others have developed.

Economics today is exploring inequalities, climate change, the future of work, financial instability, and other issues. But belief remains strong that the market should be the primary determinant of what products are produced and what prices and wages are paid. The market is also relied on as a means of addressing outstanding issues and problems. In the case of climate change, many economists advocate pricing carbon emissions or instituting a carbon tax on gasoline and other products adding to greenhouse gas emissions. Skepticism about the efficacy of discretionary government intervention is widespread among neoliberals: lower-cost policies that automatically go into effect, such as the payment of unemployment compensation, are typically favored over programs that require on-going oversight and recurring legislative approval.

So, What's Not to Love About Economic Theory?

A veneer of equality pervades discussions of the working of the market in neoclassical economics. Employers and employees both make free decisions—employees are free to leave if they are not satisfied with a job and employers are free to hire or fire whom they wish. Consumers are free to purchase whichever products they prefer, and their aggregate choices let producers know what products can sell.

Power differentials at the workplace and their economic consequences are not mentioned. Rather, the institutions that protect business owners and restrain workers' rights on the job are taken as givens. Their existence and content are assumed, not analyzed.

Human relationships within the firm, households, and government are not a focus of neoclassical microeconomics. Some economists even say that, more generally, the study of social relationships is not within the purview of economics, although they acknowledge that economics is built on assumptions about how humans behave—for example, that people act rationally and seek to maximize satisfaction. But human beings within organizations decide what will be produced, what the price of products will be, and what workers will be paid. How these organizations are structured affects the decisions made. And employees have little to no say in these decisions, nor are consumers directly involved. If the goal of economics includes improving the well-being of everyone in society, a more encompassing approach to the study of the economy is required.

Arnsperger and Varoufakis (2006) are among the many authors who point out the serious shortcomings of neoclassical economics that limit its useful applications. They point to flaws inherent in the methodology behind its models. These flaws include its treatment of society as an aggregation of individuals with no interdependencies, its characterization of individual behavior as instrumentally rational (aimed at achieving a goal), and its analysis of aggregate phenomena in terms of an equilibrium that is claimed (without proof) to emerge from the interactions of individual rational actors [An existing market price may be assumed to be an equilibrium price, but it is likely that it is not: there are always variables omitted from the analysis and the variables included are continually changing. Price adjustments are usually not instantaneous].

The assumptions made about human motivation, agent rationality, and properties of individual preferences eliminate the interdependence basic to relationships, apart from those tied to economic reward (see Goodin and Pettit 2000). Theories based on methodological individualism, such as those of neoclassical economics, are inherently inadequate for the study of social behavior because they do not allow for the fact that an individual's behavior is influenced by the groups with which he or she interacts. For example, humans care for one another, and often act for reasons other than personal gain.

A fundamental flaw of microeconomic theory, then, is its inadequacy for the study of inter-group and intra-group relationships, including power and caring relationships. This is widely acknowledged within the profession. The independence of the equations in standard microeconomic models cannot be maintained when interdependencies are introduced, and new mathematical forms are needed. While some interdependencies have been built into some models, much remains to be done before a new paradigm emerges. As it is still taught, the rational man on which microeconomic theory is based bears strong resemblance to a robot programmed to consider certain pre-determined variables when making economic decisions. Developing equations that model compromise or commitment among such robot-like creatures is an unfinished task.

Measuring Well-Being—A Tricky Problem

Economic policies are generally intended to increase well-being. But how should well-being be measured? Does a higher level of income mean greater well-being? Does longer life expectancy? More years of education? How well-being is measured matters. Welfare economics is the branch of microeconomics focused on issues related to assessing human well-being. Traditionally, total utility or satisfaction achieved by an individual provides a measure of his/her well-being. But satisfaction cannot be directly measured, and so economic indicators are used. To compare the well-being of people in different countries, the indicator most frequently used has been per capita income. Per capita income, it is argued, is an empirical measure that tracks well with the average person's access to the goods and services that yield satisfaction. One problem with this measure is the fact that income varies greatly among individuals in an area. And income does not capture important contributors to well-being, such as freedom, health, or social standing. Slaves in the U.S. pre-Civil War had higher consumption levels than industrial workers in the U.K., but lacked freedom and basic human rights—they and their children could be bought and sold, physically abused and even killed with little consequence. Freedom and rights, then, are clearly important to well-being, as are education, political voice, and more. An individual's income, then, is not an adequate measure of that person's level of well-being.

We noted earlier that a wide range of indicators of human well-being are being calculated by groups such as the U.N. Later we ask an even more challenging question: what is required for human beings to thrive? Non-market goods like a healthy and beautiful natural environment, equitable laws, caring communities and the ability to participate in public decision-making all contribute to human thriving. Many of these goods depend on the quality of government and other social institutions. One conclusion is that, in addition to evaluating the well-being of individuals, there needs to be a means of evaluating the functioning of institutions.

Market Mentality, the Pursuit of Profit and Social Ethics

As hinted at earlier, a market mentality leads to a view of social problems as problems that arise from inadequate or inappropriate valuation of a good, such as clean water or time. But when can the under- or over-supply of a desired good be adjusted through attaching a more appropriate monetary value to it? Crime is over-supplied. But is it the case that fines would be more effective at stopping crime than, say, imprisonment? Economic incentives do drive behavior, with both positive and negative effects, but other incentives are just as strong. Do we have the right balance? Or has the market economy distorted our perception of the choices we face?

The market impacts ethical standards and morals in several ways, and these effects have fueled critiques of the market system from the angle of morality. One problem that we just noted is that market or economic reasoning leads to approaching a problem from the viewpoint of finding appropriate financial incentives to end it. In an example cited by Sandel (2013), a daycare center used market incentives in an attempt to get parents to arrive on time to pick up their children—too many were arriving late. To solve this problem, late-arrivals were fined. The result was that more parents and guardians arrived late. "We're willing to pay the price to have more time for ourselves" was their apparent reasoning. An appeal to parents to be more considerate of the workers at the center would likely have been more effective.

Another example: in the European Union, firms can obtain permits to emit greenhouse gases to a certain limit, and are permitted to sell all or part of their unused emissions allowance to other firms. Using market reasoning, firms will weigh the cost of purchasing equipment that would reduce emissions against the cost of purchasing emissions permits from other firms. This permitting scheme worked, as it turned out, only if total emissions permitted were gradually reduced—and for this reduction to happen, the voluntary agreement of member states was needed. In other words, ethical behavior was the root cause of any success in reducing emissions. Pricing emissions, then, can be seen as simply a useful tool for groups who are already morally committed to the goal. And other policy tools not tied to market prices could be even more effective—such as setting timetables for transitioning to clean technologies available.

Using economic incentives to influence human behavior has its limitations in other ways. Sandel (2013) notes that virtues like altruism, generosity, solidarity, and civic spirit do not come from expected monetary rewards. "Putting a price on every human activity erodes (the value accorded) moral and civic goods worth caring about," he says (Sandel 2013: 121).

The market puts a price on goods that are limited in supply and typically depleted with use. Sandel argues that virtues are not depleted with use, but rather are strengthened and increased. The incentive for being respectful of others' time or well-being is not, and should not be, a pecuniary one. Economists should recognize this when providing guidance as to where markets do or do not serve the public good, he states.

Bruni and Sugden (2013: 161), in a companion article, state that "the telos of the market is mutual benefit." That is, the market has the effect of showing us how best "to be useful to one another" [the authors borrowed this phrase from Genovesi, who wrote it in 1766]. Through the market, we provide one another with goods and services that are needed and desired. People buying or selling are treated in much the same way regardless of social background, race, or ethnicity. The market, then, is an impartial tool and market ethics, they argue, are to be defended.

These two articles do not address inequities fostered by market institutions, even though a common observation is that the market rewards the greed that drives an excessive desire for higher profits at the expense of people and planet. Business owners can often increase profits through reducing expenditures on employees or on environmental protection measures, and the temptation to do so is strong. The consequences are apparent: appeals to businesses to show greater moral responsibility toward customers, employees and planet have had unsatisfactory effect on product safety, greenhouse gas emissions, paid parental leave, and the like. The force of moral suasion is limited.

Economists devote scant attention to the decision-making processes that lead to undesirable outcomes including rising inequality, gun violence, and environmental degradation. Most economists would be hard-pressed to offer opinions about how a different structuring of decision-making power could build stronger, more caring communities while protecting the environment and reducing inequality.

Economic Theory and Ideology: Market Myths

Although there are many branches in economics, the microeconomic theory of market equilibrium encapsulates the argument used most often by those opposing a government intervention they dislike. Government interference, they claim, hurts efficiency and harms businesses and consumers. This argument rests on a structure so flimsy it should collapse under the weight of the conclusions it supports! It is an ideological argument used to justify opinions not at all related to the efficient functioning of the market as understood in economics. As we have seen, there are many situations in which market outcomes have undesirable consequences that must be addressed. Even opponents of government intervention in fact selectively defend government intervention: they want enforcement of contracts, provision of national defense, and more but seldom want regulations that might constrain their choices.

When is it best to rely on the market, then, and when is it best to intervene, and how? There are several functions the market performs well, others where it falls short and many where unintended consequences must be handled. Opinions differ about the role that the market should play in ensuring our well-being and happiness. In looking at some market myths, we can see where an ideology justifying certain behaviors is at work, not economic facts and tested theories. Some common, but harmful, myths about the market that cannot be squared with basic microeconomic theory are listed below. Later, we consider how some shortcomings of the market can only be addressed through institutional change.

Eight Market Myths

Myth #1: Markets can only work well in a capitalist economy.

Most economies over the centuries have traded goods and services; over time, money of various sorts has been used to expedite these transactions. Today, most market economies are capitalist or are dominated by capitalists (in China, Saudi Arabia, and a few other countries, the state dominates some markets). Markets under capitalism are characterized by the ownership of businesses (including the means of production) by individuals, termed capitalists in the case of big business. Mixed economies in which there is both private and public ownership of capital are common, but in the U.S. and other democracies private ownership dominates. The mix of public and private ownership leads to various labels. Consider the case of Norway. Norway is a mixed economy and the government is the largest owner of capital. It has a democratic political system and a strong market economy. It is often called a capitalist country with a strong welfare state. It is also called a democratic socialist country.

The term market economy is often used interchangeably with capitalist economy. But markets and market competition can exist under several different ownership regimes. If all major businesses were to become nonprofits or cooperatively owned and managed, the economy would not be capitalist, but it might still be called a market economy

Myth #2: Markets are efficient.

The market provides a price mechanism for balancing supply and demand. Prices are important in distributing goods that are bought and sold; the resulting outcomes work well for persons with adequate income. Markets, though, do not ensure that public goods will be supplied or that basic research will receive funding or that jobs will pay a living wage or that children and elders will receive adequate care. The market is limited in what it can do; and its efficiency is also limited.

Markets, according to neoclassical theory, are only efficient IF perfect competition prevails. Under perfect competition, markets provide consumers with goods they want and can afford, at prices that yield only a normal rate of return above cost for producers. And even under this surreal perfect competition, efficiency is not expected to prevail in labor markets. This is because workers are not commodities produced in factories; the price paid to workers does not need to cover the cost of their 'production.' The investments that go into having, raising, and educating a child are considerable but this investment cost need not be covered by employers. Employers only need to pay a wage high enough to ensure their employees will continue to work for them. Above and beyond that, marketplace competition can raise or lower the wage paid. Note also that products produced reflect the skewed distribution of income. This is efficient in market terms, but this is a very limited efficiency. If incomes are distributed very unequally, the demand for low-cost housing and other necessities may not be met, given that what is affordable to low-income workers offers little profit in comparison to other opportunities; the demand for up-scale housing, however, will likely be met. Market efficiency does not ensure that every worker receives a living wage, let alone health insurance or a means of financing college courses. Non-market goods cannot be efficiently supplied by markets, but can be even more valuable: a healthy and attractive environment; time to enjoy friends and family; and a government that protects one's life, liberty, and pursuit of happiness are among the non-market goods that are critical to well-being.

Market Myth #3: Consumers are king; consumer choices rule.

It is true that firms respond to consumer demand for a good or service that is already being offered. But firms are the ones who decide which goods and services they will be offering. Consumers do not often get full information about the cost, safety, and quality of the goods being sold, nor are they informed of the work conditions and pay of workers producing the product or services. Hard for consumers to be 'king' under such circumstances.

Where can one demand less plastic packaging? Less use of slave labor to pick the cocoa? Products that don't suffer from planned obsolescence?

Market Myth #4: Government screws up markets.

We hear that intervention by the government leads to inefficiency and distortions in markets. This myth is particularly dangerous. A democracy in which people cannot institute laws regulating market transactions would not be a democracy. Everyone, including business people, recognizes the need for legal and police systems that protect sellers, buyers, workers, property, the value of the U.S. dollar and more. Laws are passed by Congress in response to a perceived need and, as written, are the result of compromise reached among politicians, who as we saw in Chapter 3, are especially influenced by their upper-income constituents and lobbyists. These laws are imperfect but to say that government intervention is bad is to say that we don't want our elected officials to attempt to alleviate market-related problems, no matter how dire. Opponents of government intervention, in fact, are very selective about the kinds of intervention they oppose.

Market Myth #5: The market rewards hard work and productivity.

It may, but there's no guarantee that it will. Employers are slow to raise wages as output per worker hour rises. Data show that although worker productivity has more than doubled since 1980, real weekly earnings were lower in 2019 than in 1976.[6]

Firms may find that producing more output at the same or lower cost does not increase profits if other firms are also expanding output: customers aren't always willing to buy more of a product at the same or even lower price. Farmers know only too well that a good harvest doesn't bring increased revenue if there is no demand for the increase. The COVID-19 pandemic provides a case in point: the drop in demand during the pandemic led many farmers to destroy or donate crops they couldn't afford to store or bring to market.

Market Myth #6: The market wage is a fair wage.

The simple graph of demand and supply is useful in making the point that the wage a worker is paid depends on the demand for his or her labor. When there is an economic recession, the unemployment rate rises and, for every job available, there are typically two or three unemployed workers. Supply exceeds demand at a fairly low wage level. Another point: wages are not tied to productivity or skill levels except indirectly through demand and supply. When average skill levels rise, wages may remain the same despite the increase in skills. CEO pay, also, is not strongly tied to performance but in this case the CEO often benefits; business culture plays a large role.

Market Myth #7: Capitalism drives scientific discovery.

Capitalism does foster innovation, but the goal is innovation that is profitable. Basic research that probes the origin of the universe or the intricacies of biological processes seldom results directly in a product that can be sold. The important discoveries that revolutionize day-to-day life are built on the findings of basic research that is largely funded today by the government or nonprofit foundations. Competition under capitalism does encourage applied research and the development of new products, especially ones that can be protected by patents or copyright laws. But competition can also squelch innovation. Firms prefer to avoid competition as possible. Large firms at times protect themselves from innovative competitors by buying them out, even if they have no intention of producing the competitors' products.

In addition, it should be noted that innovation is not confined to profit-seeking firms. Cooperatives, nonprofit enterprises, and other bodies also innovate and compete.

Myth #8: Competitive market pressures can end wage discrimination.

In theory, employers will hire the most productive worker, regardless of sex, skin color, sexual preference, religion, or other attributes. High output per hour is all that should matter under perfect competition. Of course, in fact, employers behave quite differently; they have strong preferences about hiring women for certain jobs, men for others, and race and other characteristics also strongly impact their choices. These employer preferences, which differ by country, are reflected in data on the distribution of jobs by race, sex, and other characteristics. Employers are well aware of the benefits they derive from the lower wages associated with occupational segregation in the workforce. The distribution of jobs by worker characteristics shows how far today's economy is removed from perfect market competition (and from equal opportunities for all).[7]

On the other hand, individuals who are able to set up their own businesses may be able to compete effectively regardless of their race, sex, and other characteristics. Getting started in a business is more difficult for minorities and women, though, because it is harder for them to get the funds needed, whether from savings, relatives, or banks. There are often other disadvantages to be overcome as well: personal networks may be less influential, for example.

Conservative and Liberal Capitalist Ideologies

The laws and customs that regulate markets vary across time and space. To countries holding colonies in the early 1900s, imperialism was beneficial; to the colonies themselves, imperialism was oppressive. The context in which markets function, then, has tremendous impact on market outcomes. Discussions of the benefits of markets, and capitalist markets in particular, are often biased—more ideological than analytical. Actual conditions on the ground may be ignored. The positive contributions of regulation or other interventions that reduce abuses and so safeguard people and the environment may be under-rated or down-played.

A belief in market economies and capitalism as the best alternative for society is quite compatible with a belief in some kinds of government intervention. There are a range of government interventions embraced by those believing in free markets, with some policies more favored by liberals and others by conservatives.

Conservative ideologies at one extreme argue against most government interventions aimed at helping individuals, such as low-income single parents with children, the long-term unemployed, those without health insurance, and more. The market, these conservatives argue, should be providing the jobs and the affordable health care and child care people need. But these same ideologues will use their belief in private enterprise to justify demands for government interventions to enable their industries to compete successfully with foreign-owned firms. They ask the government to prop up domestic demand during times of recession, to subsidize investments of all sorts, to promote U.S. products in foreign countries, to protect their investments abroad, and more. At the same time, they tend to favor minimal regulation of workplace safety and minimal requirements regarding the environmental impact of industrial activities. In brief, conservative market ideology can justify all manner of government interventions benefiting business, while opposing interventions helping the disadvantaged.

Market ideology also provides an argument for income inequality, which is claimed to be the natural outcome of market competition.

The **neoliberal policies** associated with conservative economist Milton Friedman and the Chicago school have been a major force behind the trickle-down theories justifying tax cuts for business and for the wealthy, austerity programs in countries in debt to international lenders, and more. The Heritage Foundation, a conservative think tank, reflects this mindset. It bases one component of its index measuring freedom on the claim that the less government spending, the greater citizens' freedom. The U.S. ranking in 2011 was hurt because the government increased its spending on health care![8]

More moderate/liberal supporters of the market economy favor government interventions that would reduce inequality, protect the environment, improve access to health care and more generally aid struggling families and individuals. They support programs for youth like Head Start, retraining programs for workers affected by mass layoffs and more. Still, the levels of funding they favor for such programs typically fail to adequately address the need. They are reluctant to increase taxes on the wealthy, and they support many tax breaks for business, like the accelerated depreciation allowance. They do not push for changes in business ownership rights. Both liberals and conservatives who support private enterprise want policies that enable businesses to do well.

Government as Seen in Subfields of Economics

While supportive of private enterprise, different subfields of economics have different perspectives on what government interventions are appropriate. Public finance sometimes presents the role of government as one that should ideally be derived from microeconomic theory: if only it were possible, government would determine the demand of each citizen for each public good and service, as well as each citizen's willingness to pay for those goods and services. Based on this information, the government would act as a neutral clearinghouse, ensuring that the goods demanded by citizens are provided in appropriate quantities. Macroeconomists and development economists similarly present a disinterested, unbiased government as the ideal. This government would always act in the public interest; it would see its job as promoting economic stability and growth through responsible oversight of the monetary system and the government budget.

Whatever their specialization, most economists take a pragmatic approach when it comes to the policies they recommend to officials.

They accept that the government is what it is, and present policy options intended to achieve improved outcomes without questioning how the government is structured or who wields decision-making power.

Free markets versus democratically controlled markets The liberal-conservative divide among economists extends to different attitudes toward the desirability of deeper democracy. Markets under capitalism have positive attributes. They typically reward increases in efficiency and provide a wide choice of consumer products; they also offer opportunities for new businesses, encouraging entrepreneurship and rewarding innovation and initiative. These market ideals frequently guide policy-making, with many policy wonks offering recommendations to bring market outcomes more in line with the ideals of efficiency, opportunity, and innovation.

But proponents of free markets often see improving wages and working conditions as hurting employers' bottom lines, which they oppose. In fact, higher wages for workers do drive up the labor costs of a business in the absence of increased output per worker. This basic worker/management divide is a recurring theme in arguments about government regulation of business. But would higher incomes for workers hurt the economy? The answer, as elaborated below and in Chapter 6, is no—an increase in the share of national income going to workers would shift the composition of production toward goods purchased by middle America. This could be a strong plus for the economy.

Greater political power for the middle class would strengthen our democracy and would mean stronger policies to protect the interests of working Americans. Changes in taxes, regulations, and spending could reduce business profits. This is a problem for some believers in the market, who would favor less democracy if a drop in profits would be the result. Profits, they believe, are key to people's drive to start, improve, and expand their businesses. Business could not pay higher wages and benefits, or higher taxes and remain competitive in the global market, they claim. The ideological belief in the market, then, is closely connected to the interests of business and the capitalist class in maintaining and expanding their share of total income.

Is it true that greater responsiveness to voter demands and worker needs could cripple our economy? Is it true that reduced returns to shareholders and lower CEO pay would drive investors to put their money elsewhere? The answer, in brief, is 'no,' if appropriate policies are followed. But the ramifications of major economic shifts do need to be considered in some detail.

It is necessary to identify options available to achieve the goals favored by the middle class in terms of access to health care, child care, college education, and more. The experience of Sweden and other countries shows the positive results of expanded benefits for the population. Several Swedish businesspeople have expressed the opinion that the security provided by universal health care, dependable unemployment benefits, and other programs encouraged them to undertake the risk of starting a new business. The added economic security enhanced their ability to start, improve, and expand their businesses. Sweden, by the way, had the highest number of unicorns (start-up firms that quickly became billion-dollar companies) outside of Silicon Valley in 2018 and was ranked the second most innovative country in the world after South Korea that year.[9] We later consider what would likely happen when goals other than profit become the drivers of business activity. Firms that act to provide their products while protecting the public interest can survive and thrive in a competitive market system, but there are challenges.

Government: The Good Guys or the Bad Guys?

Opponents of government intervention offer another reason for their position in addition to their belief that government has a negative impact on private enterprise. They argue that government bureaucrats and politicians pursue wasteful spending on unneeded and unsuccessful programs. They grant that some government programs are useful, but prefer that those programs be funded according to formulas that are less subject to political winds of change. For example, unemployment benefits are automatically triggered based on established rules, as are social security benefits and other payments to states and individuals. Programs that must be voted up or down periodically are more likely to be modified by politicians seeking to gain favor. James Buchanan (1986) is among those who think that efforts to make Congress more representative and accountable would be unsuccessful in overcoming the drag on the economy that bureaucracy causes. A leaner, less intrusive government would better benefit the country.

To over-simplify on-going debates, some economists treat those in government as the good guys whose laws and regulations improve outcomes, while others regard them as the bad guys. Ayn Rand's work,

Capitalism the Unknown Ideal and in particular the essay in that book on "America's Persecuted Minority: Big Business," is often cited by conservatives who view government as a necessary evil that must be contained. In that essay, Rand (1988) articulates the position of those who link the market to freedom and government to threats to that freedom:

> The mechanism of a free market reflects and sums up all the economic choices and decisions made by all the participants. Now let me define the difference between economic power and political power: economic power is exercised by means of a positive, by offering men a reward, an incentive, a payment, a value; political power is exercised by means of a negative, by the threat of punishment, injury, imprisonment, destruction.

But what about the argument that the government represents the will of the people, and therefore its proper role is to act in the best interest of the long-term economic well-being of the nation as a whole? If this is the proper role for government, the question then becomes to what extent does, or can, the government in fact represent the best interests of the people?

Here data on elections, politicians, and political campaigns come into play. As we have seen, political representation is unequal: groups with greater wealth have disproportionate voice within government. They have notable success in influencing legislation. Campaign reforms and other changes that improved democracy would shift national priorities and improve outcomes for those benefiting from increased political clout, namely those who are not among the wealthy. With greater clout, the laws and regulations passed would be more in line with the preferences of mainstream Americans (We saw in Chapter 4 that mainstream opinion has little influence on Congressional votes in comparison to the influence of business lobbyists).

Government can in fact better represent the will of the people. Blacks, women, the disabled, the LGBTQ+ community, and others have had their lives improved when they succeeded in advancing their causes. People are making their dissatisfaction with the current state of affairs clear; they want an improved quality of life—markets that function more smoothly matter only to the extent that they lead to greater well-being. (Those outside the U.S. also deserve representation in U.S. policy debates that greatly affect them.)

The failure of economic theory to analyze power makes it difficult to answer the question of how to institute greater democratic control of policymaking in a way that will improve outcomes. Certainly, citizen input on proposed legislation, and citizens' ability to propose legislation, can and should be made more impactful.

Power Neglected, Right and Left

Neither conservative nor liberal economists have done much empirical research on the exercise of power by government and the powerful. Few have studied how greater input into policy decisions by those affected can improve outcomes, for example. The typical approach to political power seems to be: government functions the way it functions; it is what it is. Our role as economists is to make recommendations that will improve outcomes, not to recommend restructuring the nation's political machinery. The institutional frameworks that support authorities, however, change over time as we have seen. Changed understandings about the world change the limitations placed on authority.

Research that doesn't explicitly recognize which groups hold power in government and how they exercise it is research that will be biased toward maintaining the status quo. Voices funded by the wealthy have dominated public discourse for decades and the alternative perspectives of those who push back against the policies of the powerful receive far too little recognition. However, there are always political struggles and so an ever-shifting balance of power that deserves to be considered by researchers.

Economists typically do not explicitly acknowledge this shifting balance of power. Rather, when they make policy recommendations, they judge what policy options seem feasible under prevailing circumstances and only present those options. Analysis of the power dynamics which has narrowed feasible options is left to political scientists.

In some subfields of the profession, however, the situation is a bit different. Development economists have long recommended institutional changes to developing nations. Other subfields of economics also address power-related questions, including feminist economics, Marxist economics, labor economics, economic history, and others. These subfields draw heavily upon interdisciplinary insights.

Karl Marx is perhaps the best-known thinker within these alternative subfields. The imbalance in power between the capitalist business owner and his employees was basic to his analysis of the capitalist system. There are other well-known economic theorists who have devoted considerable attention to the inter-connections between power and the economy. John Kenneth Galbraith, a Harvard economics professor and advisor to President Kennedy, is an example of an economist who wrote on power. In his book, *Anatomy of Power* (1983: 4), he lamented "It is a measure of how slightly the subject of power has been analyzed that the three reasonably obvious instruments of its exercise do not have generally accepted names." The three are, according to Galbraith, power based on rewards, based on penalties, and based on shared values.

He was also concerned that economics as taught in introductory classes serves the purpose of protecting the power of big business by spreading the ideology of the market. This ideology, as summarized by Ayn Rand (1988), is as follows: "In a free market, all prices, wages, and profits are determined—not by the arbitrary whim of the rich or of the poor, not by anyone's 'greed' or by anyone's need—but by the law of supply and demand." The mechanism of a free market reflects and sums up all the economic choices made by all the participants. People trade their goods or services by "mutual consent to mutual advantage, according to their own independent, uncoerced judgment." A person grows wealthy through offering better values—better products or services, at a lower price—than others. No exploitation or unfairness is acknowledged in this view of the capitalist market as driven by the impersonal forces of demand and supply.

Warned Galbraith (1983: 120): "Nothing is so important in the defense of the modern corporation as the argument that its power does not exist—that all power is surrendered to the impersonal play of the market, all decision is in response to the instruction of the market."

This message in a slightly different form is being spread today by the Council for Economic Education, which aims to impress upon children at all grade levels, from kindergarten through high school, the blessings of living in a capitalist economy. Their focus is on personal finance, and positive messages abound about moving ahead through hard work and saving. You, too, can become a millionaire! Any discussion of systemic problems with the economy would blunt the positive spin they put on today's realities. They aim to get economic education classes in high schools and grade schools across the country, and offer a broad range of materials and resources in support of their message.

Kenneth Boulding, former president of the prestigious American Economic Association, also offered a perspective on power in his 1990 book, *Three Faces of Power*. Although there are parallels to Galbraith's three forms of power (power based on penalties, rewards, or persuasion), Boulding focused more on the effects of power: destructive, productive, and integrative. While he acknowledged that the exercise of power is critical to economic outcomes, he stopped short of arguing for the integration of power into basic micro- and macroeconomic theory. Eric Schutz (2011), in a detailed critique of mainstream economics, has shown that underlying social realities of class and power are essential in explaining economic inequality and that their neglect in economic theory biases policy prescriptions.

A Nobel laureate in economics, Amartya Sen, is more typical of prominent economists in that he has not classified or analyzed different kinds of power, despite being familiar with the various debates and questions raised. As he explained (Sen 2004: 353), "Perhaps the point is not whether the concept of power is being used (by me), but about the frequency with which the word 'power' occurs in my writing…as opposed to the ones I tend more often to use, such as 'empowerment' or 'capability' or 'freedom' or 'agency' or "threat' or "vulnerability."

He refers to his article "Women's Agency and Social Change" in which he argues for "adequate recognition of political, economic and social participation and leadership of women" (Sen 1999: 203). In that article, he also speaks for increasing women's earning power, economic role outside the family, literacy, education, and property rights (Sen 1999: 189–203). This is an approach to empowerment that focuses not on the process of institutional change, but rather on policy goals that would result in greater power and opportunity for individual women. Such policy goals cannot be achieved without political backing, however. Collective efforts of self-organized women are required to advance such agendas—that is, the collective exercise of power by women.

Alternative Perspectives: Marxists, Feminists, and Other Heterodox Economists

Feminist, Marxist, ecological, behavioral, institutional, and other economic theorists bring in-depth insight into the systemic problems of an economy whose driving force is monetary gain. Their critiques of mainstream economic theories point toward alternative ways of valuing people and their work, and the gifts of the natural world. They offer alternative understandings of the motivations behind human behavior and of the factors involved in institutional change.

Alternative approaches to the economy bring in knowledge from other academic disciplines; most clearly recognize the embeddedness of the economy in the natural world. Society depends not only on the economy in the narrowest sense—the production of goods and services—but also on the resources of the natural world and on all kinds of social behaviors. Society requires the reproduction of human life, which involves the imparting and expansion of knowledge and skills. Society depends on the institutions embodying collective agreements that shape how activity is organized. It depends on caring relationships. Heterodox economists focus on different aspects of the interdependencies that are basic to the functioning of business, households, and government. Institutional economists, ecological economists, and other scholars in non-traditional fields of economics are bringing new aspects of valued goods and actions to the fore and so are reshaping thought about how to approach living and working together. The goal of policymakers should be to improve the well-being of both people and planet through changing behaviors—by relying on different sources of motivation.

Marxists and radical economists have a focus on inequalities in power at the workplace, and the contradictions within the capitalist system that lead to economic crises, division and repugnant inequalities. They see the achievements of capitalism as based on exploitation, and understand that if working people recognize this, they can band together and establish a more democratic, egalitarian society.[10] Feminists stress the complexities of women's position, examining women's paid and unpaid work, and the division of labor within the household and broader society as shaped by patriarchy. The undervaluation of caring labor is a key concern, and many feminists are advocating that care—for one another and for the planet—be a core organizing principle of social activity. Environmental economists stress the value of non-market goods, like clean air and the preservation of species and ecosystems. Institutional economists examine the "social contracts" that hold society together, as embodied in institutions. Behavioral economists note the predictable ways in which human behavior is not rational, and suggest better ways of understanding how and why people behave as they do. There is also a growing literature on solidarity economics (see Benner and Pastor 2021), which focuses on cooperative and mutually beneficial ways of providing needed goods and services.[11]

Feminist Economics and the Complexities of Human Motivation
Feminist economists are all aware of the difference in status of men compared to women. Women's lived experiences differ markedly from those of men, in the home and in broader society. Over the years, women's legal rights have been restricted and remain so, despite some improvement. In the U.S., customs regarding the division of labor between women and men continue to affect households, firms and the judicial system. Institutionalized practices are a key element in explaining women's income and occupations.

Women do most of society's caring labor, which includes unpaid work in the home. This care work is under-valued: it is typically low-paying or unpaid despite the level of skill and effort often demanded. Unpaid labor is not included in gross domestic product (GDP) and other measures of output. Yet in developing countries, unpaid labor may account for 40% or more of a nation's production.[12] Even in advanced countries, unpaid household and volunteer labor are critical to maintaining living standards.

People, male and female, do not, in fact, act like the utility-maximizing machines of microeconomic theory. Women and men often make their labor market decisions, not to maximize their lifetime earnings flow or similar measures of self-interest, but instead balance future potential earnings with the needs of their families and their desire to contribute to their communities. Consumer purchasing decisions similarly take into account household as well as personal needs over time. For too long, microeconomic theory treated the decisions of women or men made for the purpose of improving the well-being of others, such as one's family, as irrational: by definition, the rational decision is the one maximizing one's personal good! Feminist economics, then, questions the assumptions on which standard economic theories have been based. It imagines a society based on care in which all can flourish. It is re-imagining what the variables and values are that in fact explain human economic activities[13]

Concluding Remarks

Economic theories lie behind the rules of the economic "game" crafted by lawmakers. They also shape the economic policies designed to help businesses and households, and to solve national problems. However, these theories traditionally have not taken into account the impact of human interdependencies on economic behavior. The complexity of human motivations has been ignored, reduced to the pursuit of satisfaction, with no room for love or altruism. More recently, many fields of economics have modified traditional theories by incorporating a richer understanding of human behavior. However, the ideological belief in the market economy remains solid. Power relationships are still considered to be beyond the scope of economics, and are set aside for study by political scientists and philosophers. Yet economic policies that benefit people and planet depend on a government that is responsive to the needs of people and planet. Changing how power—economic and political—is structured is basic to addressing today's crises. Rethinking how power is structured in our democracy, then, is urgent. The next chapter addresses the following: How can we achieve a more responsive and responsible government? What policies would enable us to hold big business accountable for harms inflicted? Can the composition of corporate boards be changed in ways that would prevent business from engaging in activities that have lamentable or even catastrophic consequences?

Appendix: Shortcomings of Typical Economic Analyses

The Failed Response to Katrina and the Financial Crisis

An examination of the shortcomings of a typical economic analysis of these failures.

The Katrina crisis and the global financial crisis were each marked by failures of the market and of government. The kinds of policy recommendations coming from traditional economic analyses are discussed below, along with critiques of such traditional approaches.

Failures in New Orleans. In 2005, before the hurricane hit, New Orleans was a city with widespread, persistent poverty: 24% of its inhabitants were poor. The poverty of most of the stranded survivors is critical to explaining why they were left abandoned for so long after the storm: along with poverty comes a lack of political clout. Racism was also a major factor.

Poverty that persists over time usually indicates a market failure: the operation of supply and demand in the labor market cannot be expected to overcome employment problems caused by poor education, disability, and caregiver responsibilities. The poverty rate in New Orleans, despite the loss of thousands of its poorest citizens, was back to the pre-Katrina level of 24% by 2010, and rose for several years after. The reductions in low-cost housing, public health care and transportation since Katrina contributed to this rise in the poverty rate. By contrast, the U.S. poverty rate has been notably lower, at 13% in 2005, 15% in 2010, and 12% in 2020. To end persistent poverty, government should ensure that the root causes of poverty are addressed. The poor need access to training and to services like child care and transportation before they can get a job that pays a living wage. This market failure, then, requires government intervention, but here the government failed as well.[14]

Flood protection infrastructure is a public good and inhabitants of New Orleans need an adequate levee system. Several governmental bodies share responsibility for ensuring New Orleans is protected. The government's failure to fulfill that responsibility indicates to economists a need for reforms that would improve the facts provided by government.

There are several tools that economists use when approaching this task. Based on information they receive from relevant agencies and industries, they can present estimates of the cost–benefit ratios of alternative public investments. They evaluate alternative means of raising adequate revenues and can suggest methods for determining the spending preferences of residents. Vigdor (2008) offers an assessment of the impact of Katrina in the highly-respected *Journal of Economic Literature* that makes good use of traditional tools.

But the decision about which option to choose is treated as a political question. The problem of ensuring that the political process behind government and voter decision-making is up to the job is one left to political science researchers, largely in academia, or to government itself. Yet, the line between politics and economics is routinely crossed. Economists and other policy researchers formulate their recommendations with an end user in mind—be it a particular political party or city agency. Most typically, they focus on developing recommendations that the leadership currently in place might adopt and implement in the short term. If long-term planning is demanded by the problem, a task force or committee will be set up to study the question. Recommendations, then, are issue-directed within the current context and seldom aim at improving who makes decisions and how. The status quo becomes further entrenched.

Should economists or other policymakers wish to improve the policy decision-making process, they are handicapped by economic theories and models that are weak in their description of the behavior of the leadership of large organizations. Empirical research on the critical factors that influence leaders in setting priorities, or voters in their electoral decisions, is relatively sparse. For example, much of the work that has linked income and class to voting records and policy preferences statistically has been done since 2000. As recently as 1965, scholars could argue that the multitude of interest groups lobbying Congress provided a reasonable balance among the different sectors of the economy, thus keeping the legislative process unbiased (Olson 1965).

The more basic positive question of whether the city will come back, however, is fundamentally an economic one. After Hurricane Katrina, will the city of New Orleans continue to be a preferred location for more than 400,000 residents and their employers? Or will the disaster shift the city to a new equilibrium level of employment and population?

It is taken for granted that the powerlessness of the average citizen of New Orleans will continue. As one economist put it, Can New Orleans be rebuilt is fundamentally an economic question (JEP December 2008). He argued that "But the question of whether New Orleans will come back is not primarily one to be decided by market forces. It is a question challenging New Orleanians as they plan their future. Isn't this really a question of what New Orleanians want for their city, given political and economic realities? Framing it as an economic question takes away agency from the people who make up the city—implying that outside forces will shape their destiny. Although the city may lack a history of successful entrepreneurship in today's leading industries, it is premature to assume that this past alone will determine its future. If policy makers were to agree on a new direction, supported by grassroots, a new course could become a possibility.

The financial crisis played out in the context of extensive government involvement in the financial sector. A stable monetary system is a highly valued public good, and government oversight of the money supply and banking system is essential. The strength of the U.S. dollar rests on the fact that it is backed by the full faith and credit of the federal government. As part of its role in maintaining the integrity of the dollar, the Treasury imposes several regulations on banks. These regulations are designed to ensure that loans have adequate backing, that each bank maintains adequate liquidity, and that the risk in bank portfolios is at an acceptable level.

In addition, financial markets are not perfectly competitive. As economists use the term, a "perfectly competitive" market has several characteristics. Products in such markets are clearly understood and easily compared. Rates or prices are determined through competition, so that the price of a good is close to its cost, and the consumer faces several suppliers offering identical products. This is not the case in the financial sector. Instead, there are ways for a firm to gain an advantage over its competitors. In fact, pressures of competition lead banks and insurance companies to find ways of gaining advantage. For example, a bank may differentiate its product through advertising and location, through special deals on a package of products and services. A bank may have a local monopoly in the sense that it is the only bank in town. Consumers often face a difficult and time-consuming process when collecting the information needed to make a sound financial decision regarding loans or the purchase of insurance or other financial assets, such as mutual funds. Information comparing the alternative charges, locations, interest rates, risk, and rates of return of different products may be limited or incomplete. This imperfect information means that fingers are crossed when choosing a package of services from one company rather than another. Given these problems of market imperfections, government intervention is needed to protect consumers. Government regulations require disclosure of information regarding costs and fees associated with mortgages and other financial products, for example.

As we know, despite extensive regulation and restrictions on their activities, banks invested too much in risky assets during the build-up to the 2008 financial crisis. Risky behavior by banks has long been a recurring problem; regulators are well aware that risk entails the potential of default on obligations to account holders, with the devastating consequences that imply. The financial industry has tremendous clout, and this is reflected in the very muted reform efforts. Again, the fear of adverse consequences for taking risks, not following rules, etc. was not great, and undoubtedly contributed to the widespread nature of the problem. While government regulation has improved the safety and security of banking services, problems remain. Many economists favor stricter regulation, but only those measures embodied in Dodd-Frank achieved the consensus needed to make it through the Congressional process. Banks remain too big to fail. The most likely remedies, such as a return to Glass-Stegall or breaking up the largest banks, are vehemently opposed by large banks, who largely prevailed in the debates in Congress over the content of the 2010 Dodd-Frank financial reforms. Power struggles among many industries and organizations were critical to the outcome of these debates.

In the case of the financial crisis, as in the poor emergency response to Hurricane Katrina, the policy debates ended with changes that many would classify as marginal. In the case of New Orleans, the levee system was physically strengthened and elected levee boards were done away with. There was agreement that the wetlands should be protected, but funding provided was inadequate. The poverty rate remains high, and the housing situation would be even more dire, were it not for the work of nonprofits.

Americans for Financial Reform was the leading grassroots coalition taking on financial reform. This coalition fought long and hard for substantive reform measures. Many of their concerns, however, have not been addressed. Only a handful of bankers have been found guilty of breaking the law.

The forces that left so many with so little voice are being tackled, not by elected government officials, but by organizations that people have created themselves. Commentators and researchers who stressed the need to increase democratic, participatory control over government agencies remain difficult to find. But in fact holding these agencies more accountable to the public they serve does have the potential to make them more responsive.

Market-based analysis, such as that given above, has the effect of focusing the search for answers to problems on improving the functioning of the market. An alternative analysis that took all major factors into account would consider how government decision-making could be made more accountable to those affected by those decisions.

Notes

1. https://www.cnbc.com/2019/02/26/warren-buffett-on-wealth-inequality-us-should-take-care-of-its-own.html.
2. The criticisms of capitalism made by Smith receive less attention, but he was well aware of its negative impacts on workers and of the temptations for industrialists to influence politicians to their benefit. Michael Perelman, an economics professor at California State Univrsity at Chico, provides a nice overview at: https://www.nakedcapitalism.com/2014/06/michael-perelman-adam-smith-advocated-controls-workers.html. Excerpt Smith states that the interest of manufacturers and merchants "in any particular branch of trade or manufactures, is always in some respects different

from, and even opposite to, that of the public." "The proposal of any new law or regulation of commerce which comes from this order (employers) ought always to be listened to with great precaution, and ought never to be adopted till after having been long and carefully examined... It comes from an order of men whose interest is never exactly the same with that of the public, who have generally an interest to deceive and even to oppress the public, and who accordingly have, upon many occasions, both deceived and oppressed it" (See also Smith 1789, Book 1, Chapter 10, Part 2 where he discusses, without criticism, government policies that hold down wages).

3. https://moneyandmarkets.com/warren-buffett-on-why-companies-cannot-be-moral-arbiters/.
4. https://www.forbes.com/billionaires/#61a8669251c7.
5. https://www.cnbc.com/2019/12/31/what-billionaires-said-about-wealth-inequality-and-capitalism-in-2019.html.
6. See the Council of Economic Advisers, *Economic Report of the President 2021.* Appendix Table B-30.
7. See endnotes 9 to 14 in Chapter 4.
8. https://www.theatlantic.com/business/archive/2011/01/the-10-freest-economies-in-the-world/69611/.
9. The Scandinavian country is home to Europe's largest tech companies and its capital is second only to Silicon Valley when it comes to the number of "unicorns"—billion-dollar tech companies—that it produces per capita. https://www.weforum.org/agenda/2018/02/south-korea-and-sweden-are-the-most-innovative-countries-in-the-world/.
10. The Union for Radical Political Economics publishes a journal and provides information and resources related to radical economics at: www.urpe.org. It is a member of the Allied Social Sciences Association (ASSA) and organizes several panel sessions, often with other heterodox economists, for the annual ASSA conference.
11. Several resources are listed at: https://solidarityeconomics.org/get-informed/. The World Social Forum brings together solidarity activists from around the globe. Find out more at https://wsf2021.net/ and https://www.ussocialforum.net/. There are also dozens of solidarity networks around the world focusing on specific locations and/or specific issues.
12. https://www.ilo.org/wcmsp5/groups/public/---dgreports/---dcomm/---publ/documents/publication/wcms_633135.pdf.
13. The International Association for Feminist Economics (IAFFE) publishes a journal and gives useful insight into the scope and impact of feminist economics at its website www.iaffe.org. It is a member of the ASSA and has sessions at the annual ASSA meeting (see above endnote). Books, including Barker et al., *Liberating Economics* (2021), Collins (2000) and Bergmann, *The Economic Emergence of Women* (2005), provide analysis

of critical issues from a feminist perspective. Peterson and Lewis (1999) provide an overview of key issues.

14. https://www.brookings.edu/blog/the-avenue/2015/08/27/concentrated-poverty-in-new-orleans-10-years-after-katrina/. American Community Survey data show the changes in numbers of poor versus total population: 2005, 106.7/436.1; 2006, 47.2/212.9; 2010, 92.0/338.9; 2015, 88.9/379.7. https://talkpoverty.org/2016/08/26/people-poverty-work-paul-ryan-misunderstands-poverty/ explains impact of poverty. https://www.census.gov/quickfacts/fact/table/neworleanscitylouisiana/PST045216 for recent data.

References

Arnsperger, Christian and Yanis Varoufakis. 2006. What is Neoclassical Economics?. *Post-Autistic Economics Review* 38(1):1–10 (Available at www.paecon.net).

Barker, Drucilla, Suzanne Bergeron, and Susan Feiner. 2021. *Liberating Economics: Feminist Perspectives on Families, Work, and Globalization*, 2nd ed. Ann Arbor, MI: University of Michigan Press.

Benner, Chris and Manuel Pastor. 2021. *Solidarity Economics: Why Mutuality and Movements Matter*. New York: Wiley. Open source copy at: https://solidarityeconomics.org/se-book/.

Bergmann, Barbara R. 2005. *The Economic Emergence of Women*, 2nd ed. New York: Palgrave MacMillan.

Boulding, Kenneth. 1990. *Three Faces of Power*. Thousand Oaks, CA: Sage Publications.

Brown, Doug. 1999. Postmodernism. In *Elgar Companion to Feminist Economics*, eds. J. Peterson and M. Lewis, 628–634. Northampton, MA: Edgar Elgar.

Bruni, Luigino, and Robert Sugden. 2013. Reclaiming Virtue Ethics for Economics. *Journal of Economic Perspectives* 27 (4): 141–164.

Buchanan, James. 1986. *Liberty, Market and State: Political Economy in the 1980s*. New York: New York University Press.

Cohen, Jonathan D. 2005. The Vulcanization of the Human Brain: A Neural Perspective on Interactions Between Cognition and Emotion. *Journal of Economic Perspectives* 19 (4): 3–24.

Collins, Patricia H. 2000. *Black Feminist Thought*. New York: Routledge.

Council of Economic Advisers to the President. 2021. *Economic Report of the President*. Washington, DC: Government Printing Office. Appendix tables B-30&32. https://www.govinfo.gov/app/collection/erp.

Galbraith, John K. 1983. *The Anatomy of Power*. Boston: Houghton Mifflin.

Goodin, Robert, and Philip Pettit, eds. 2000. *A Companion to Contemporary Political Philosophy*. Malden, MA: Blackwell Press.
Keynes, John Maynard. 1936. *General Theory of Employment, Interest and Money*. London: Palgrave MacMillan.
Marx, Karl. 1848. *The Communist Manifesto*. https://www.bl.uk/learning/timeline/large106572.html.
Olson, Mancur. 1965. *The Logic of Collective Action* (rev.). Cambridge: Harvard University Press.
Page, Benjamin, Larry Bartels and Jason Seawright. 2013. Democracy and the Policy Preferences of Wealthy Americans. January 2013. A Russell Sage Foundation Report. https://www.russellsage.org/research/reports/wealthy-americans-political-preferences.
Peterson, Janice, and Margaret Lewis, eds. 1999. *Elgar Companion to Feminist Economics*. Northampton, MA: Edward Elgar Publishing.
Rand, Ayn.1988. Quotations from H. Binswanger, ed. *Ayn Rand Lexicon* 1988. Dutton Adult Publisher at: http://aynrandlexicon.com/lexicon/free_market.html.
Rorty, Amelie O. 1992. Power and Powers. In Rethinking Power, Wartenberg, Thomas, ed., 1–13. Albany: State University of New York Press.
Sandel, Michael. 2013. Market Reasoning as Moral Reasoning: Why Economists Should Re-engage with Political Philosophy. *Journal of Economic Perspectives*. 27 (4): 121–140.
Schutz, Eric A. 2011. *Inequality and Power*. New York: Routledge.
Sen, Amartya. 1999. *Development as Freedom*. New York: Random House.
———. 2004. Continuing the Conversation. In *Capabilities, Freedom and Equality*, Agarwal, Bina et al. eds. New Delhi: Oxford University Press.
Simon, Herbert. 1991. Organizations and Markets. *Journal of Economic Perspectives* 5(2):15–44.
Stiglitz, Joseph. 2020. Measuring What Matters. *Scientific American*. August 2020. 24–32.
Tversky, Amos and Daniel Kahneman.1986. Judgment Under Uncertainty. In *Judgment and Decision Making*, Hal R. Arkes and Kenneth R. Hammond, eds., 44–45. Cambridge: Cambridge University Press.
Vigdor, Jacob. 2008. The Economic Aftermath of Hurricane Katrina. *Journal of Economic Perspectives* 22(4):135–154.
Wolff, Edward. 2021. Household Wealth Trends in the United States, 1962 to 2019. 2021. Cambridge, MA: National Bureau of Economic Research. Working Paper 28383 http://www.nber.org/papers/w28383.

CHAPTER 6

Building a Caring, Solidarity Society

Advocates for a new economic order—one that is just and sustainable—can be found almost everywhere these days. Many of us, whether billionaires, cab drivers, or school teachers, are more than ready for a shake-up that brings bright new possibilities for the planet. Has the time finally arrived for a shift-change in how we do things? Unfortunately, there is little agreement about how this imagined new world would work, which makes its realization well-nigh impossible. Take the basic question of who will wield power under this transformed society. Billionaires, not surprisingly, foresee little change in how power is structured; they argue instead that a change in thinking, and a change of heart, among national leaders can lead to an eco-friendly, shared prosperity. They argue that new ways of thinking are all that is needed to bring new policies that move us toward sustainable energy sources and improved educational and job opportunities. A reformed capitalist society is their dream. They believe that the competitive drive for ever-greater wealth can be harnessed to serve people—although they cite no entity capable of doing such harnessing.

At the other end of the thought spectrum on the matter are those who demand a different distribution of power, with an end to capitalist domination of society. They envision a radically deeper democracy: the people will wield power through accountable representatives and big money will be out of politics. Workplaces will no longer be governed by a profit-seeking elite who set the rules. Caring labor, both paid and unpaid, will

M. T. Hill, *Building a Solidarity Society*,
https://doi.org/10.1007/978-3-031-07349-6_6

be valued. Mother Earth will be honored. All indications are that this more democratic vision is the only one that stands a chance of moving us toward deeply transformative goals.

In previous chapters, we have shown that capitalism is at the root of many of today's crises; that profit takes priority over people and planet for those at the top of the capitalist "game." The wealthy elite defend capitalism and capitalist values; they make concessions to urgent demands only when compelled to do so. Currently, support for capitalism as it stands is eroding, but the opposition is not sufficiently unified to move the nation forward to radical or deep democracy. Under radical democracy, meaningful reforms would take place, although capitalism would continue to dominate economic production. Even further progress could come next: wealth generated by collective effort would be collectively controlled and directed in line with the best interests of communities and the environment. Under this non-capitalist solidarity society, the focus of political effort would be on ensuring that institutions serve such values as equality, democracy, compassion, and sustainability.

There are stages, then, in the progress toward a solidarity society which would enable people and planet to thrive. In this chapter we look at what is involved in such thriving—basic principles; then we go into the specifics of how a caring solidarity society would function; and finally we address how we can move forward toward that society.

To thrive, both people and planet need to have basic needs met and more. We start our exploration of what a thriving, sustainable society should be like by considering proposals that aim to reform and reshape capitalism. We revisit the shortcomings of that approach in light of the goal of thriving. As we have seen, progressive reforms within a capitalist society are critical steps on the way to a solidarity society. But more than reform is needed. We look at some progressive reforms in the past, some that are on the table now, and then turn to the political steps needed to move beyond reform and realize the goal of a non-capitalist, liberating society.

A reformed capitalism has limited possibilities. The wealthy who support meaningful economic reforms generally are firm believers in capitalism, and they are not alone. Many of us share Warren Buffet's belief that we mustn't kill the goose that lays the golden eggs. Capitalist reformers recognize the dangers of global warming and, within the companies they own or manage, are taking steps to reduce emissions of greenhouse gases. Corporations, including giants such as Walmart and Amazon, already have policies in place that have led to meaningful reductions in emissions per unit of revenue. Such policies stress energy efficiency and more recently some firms are proposing a transition to clean energy via a switch to electric vehicles and the like. Bezos, Amazon's former CEO, declared "We want to use our scale and our scope to lead the way" to a new, sustainable economy.[1] Amazon has plans to convert its vehicle fleet to all-electric models over the coming years. Walmart, too, has ambitious goals for emission reductions. These two companies, however, have not shown similar progress in improving the lives of their workers: both are notorious for worker dissatisfaction with their labor practices and both are strongly opposed to unionization.

Besides moving toward sustainability, many business leaders across the country are taking progressive steps in other areas. Some are giving their employees shares in the company as a means of increasing both worker income and productivity. Many have made commitments to socially responsible principles, for example by signing on to the U.N.'s Environmental-Social-Governance (ESG) principles which include fair pay and sustainability. Many are donating to community charities and to groups working for social justice.

In addition to the direct efforts of individual firms, researchers in elite think tanks are advising businesses and governments that capitalism can thrive under economic models more suited to the twenty-first century. New, green technologies offer growing market possibilities while moving the nation toward net zero greenhouse gas emissions. New agricultural practices can be regenerative and resource-saving. Social inequalities can be reduced through arrangements that permit a rate of return to government on research it has funded. *Doughnut Economics (2017)* and *A Finer Future (2018)* are two books fleshing out such arguments.[2] They are authored by members of the Club of Rome, a prestigious international think tank. The authors argue that forward-thinking, courageous business men and women will be leaders in the transformation of the economy. These businesspeople have the ability to convince national leaders of the need for change and can inspire in them the commitment to sustainability and justice that is required.

These authors do not question the current rights and privileges of business owners, but argue for policies that would have a redistributive effect, especially on newly created wealth. Several of the policies they favor would be controversial and would be opposed by lobbyists for business. The authors at times acknowledge the political challenges of winning support for their more radical policies. In discussing the rise of social safety nets and worker protections in the mid-twentieth century, the authors of *A Finer Future* state: "...these shifts occurred because advocates discredited the existing story, created a new narrative, weakened existing power bases, and constructed new ones."[3] They allude to political struggles, then, but the critical importance of grassroots political mobilizations for change is masked in such vague generalizations. This treatment of political struggle is repeated whenever the difficulties involved in passing needed legislation come up.

For example, reversing the financialization of the economy that has occurred in recent years requires a reining in of the speculative risk-taking of the banking sector—something that was not achieved even after the near-collapse of global banking in 2008–2009.[4] Legislation and proposals to move us closer to sustainability and net zero greenhouse gas (GHG) emissions suffered setbacks under the Trump administration and progress under Biden has been slow relative to need (and could be reversed). As the 2022 U.N. report on climate change notes, a massive yet equitable transformation of the global economy is required to avoid environmental catastrophe, but political progress is unacceptably slow.[5] Political struggles are inevitable when the change that is required demands that sustainability and equity replace the drive for profit as the motivating force of economic decisions at all levels. Large losses for some businesses would be inevitable as they shift their products or production processes. Piketty (2022) may seem more realistic in his call for major reforms and a revitalization of social democracy, but he too devotes insufficient attention to the problem of creating the popular political mobilization required for the reforms he proposes.

In brief, no matter how compelling the arguments for new ways of organizing production, argumentation alone will not be sufficient to change long-established ways of behavior. Yet the idealistic notion that right will prevail without difficult political battles permeates much of the literature that advocates for substantial reforms within capitalism. These reformers claim that today's mess is largely due to mistaken policies that result from mistaken theories. Change those theories to ones that will bring economic and environmental health and healing, and national policies will become life-giving and transformative.

But it has always been the case that those in power only adopt policies that advance their interests—ones that are ideologically compatible with their priorities. Difficult political struggles have always been necessary to wrest their consent to substantive reforms.

Reforming capitalism, then, is a strategy limited by the political power of the reformers (Hill 2007, 2021). Gains tend to be piecemeal. In addition, any reforms set into place may be rolled back when political winds shift, as happens constantly. In Europe, while universal access to health care and low-cost higher education remain the norm in many countries, these and other programs have been subject to severe budget cuts in recent years. In the U.S., many social protections have been eroded. Pre-pandemic, fewer than half of the officially unemployed received benefits, a shocking drop in coverage from earlier years. The value of the minimum wage, which is not tied to the rate of inflation, has been falling in real terms for years. Its purchasing power in 2021 stood at about the same level as in 1979.[6] The soaring cost of a college education, even at public universities, has resulted in skyrocketing student debt; welfare reforms have meant that fewer families qualify; and so on. Gains in environmental protection are always at risk. Enforcement of environmental regulations fluctuates over time along with budgets, and regulations themselves are modified or eliminated. Regulations don't keep up with changes: newly created toxic chemicals may not be designated as hazardous for decades.

How long can we afford ups and downs in the commitment to environmental protections? Reducing man-made greenhouse gas emissions is imperative at this juncture. While planet earth may be untroubled by human travails due to the collapse of social safety nets, the rise in atmospheric temperature is another matter. Global warming is becoming more and more devastating for earth's ecosystems, which makes it imperative that profit-making be dethroned from its place of honor in the economy. Activists who believe in a reformed capitalism argue that net zero emissions can be achieved in market-driven economies. They have confirmed the financial and practical feasibility of net zero emissions, but have failed to point out a political path to this goal. How many businesses would willingly accept reduced profits and market shares? How many environmentally destructive industries would agree to change their production processes and product lines, even if they could survive the change? Up to this point those arguing for net zero emissions have lacked the clout to force their congresspersons and legislators to vote for a comprehensive package that could in fact get us to net zero! And this begs the question of consistent enforcement.

In the end, the greatest problem with a kinder, gentler reformed capitalism is that it only lasts as long as the strength of the political coalition that has forced the kindness upon it. And at this point we don't have the luxury of vacillating in our response to climate change. Reformers would do well to go for a lasting reformation—of the sort that ends capitalist domination of production and political policies. A coalition powerful enough to end climate change can also end the domination of our democracy by the powerful wealthy. In fact, the two goals are inseparable. Ending climate change requires democratic control of big businesses. Every big business should be answerable to a board that acts in the best long-term interests of the country and the planet.

Change that is seen as critical and urgent puts the notion of what is feasible in a different light. Once a new political formation becomes dominant—one that has an ideology of sustainability and justice—we would be able to move ahead with putting needed policies in place, without having to concentrate energies on fighting defensive battles to safeguard past gains.

Thriving Requires the Basics and More

Taming capitalism is a necessary stepping stone on the path to transforming society, but if we are to safeguard the planet and its people in the long term, the end goal must be a society based on the principles of solidarity and care. In such a society, policies will aim to establish an egalitarian democracy in which people and planet thrive. There are principles to guide us on our way. In this section, we consider these principles, including some that have guided global efforts to improve lives both in the past and at present. Then in the next section (The Path to a Thriving Society), we look at policies and political changes that would in fact bring us closer to realizing the vision of a thriving society—policies that have already had some success. In the third section (Getting There from Here) and the final section (Happening Now and Next Steps Toward Solidarity), we consider where we stand now on the path to a flourishing solidarity society and the next steps that would move us ahead.

Thriving requires first that basic needs be met. A liberating, solidarity society, then, will strive to ensure that basic needs are met, including people's needs and the needs of the earth's many species and of the planet itself. Beyond this, individuals thrive when they are able to function to their fullest capabilities, and the planet will thrive when its damaged ecosystems are restored to full health. Fulfilling all these goals may not be a realistic possibility, but the society for which we aim—a solidarity society—would move with strength and will in this direction.

Guiding principles. What does it take to meet basic human needs? And, beyond the basics, what is required for people to live rich, full lives? To answer these questions, we begin with a compacted version of Maslow's list of human needs. His list, first published in 1943, was not intended to guide social policy. Since then, scholars have analyzed what is required for people to thrive and how policy can move a country in that direction. The capabilities approach, pioneered by Amartya Sen and providing the theoretical grounding for the U.N.'s Human Development Reports, clarifies what is needed: a human being thrives when he or she is able to fully use all his/her capabilities. This requires that basic needs be met, but more as well. Martha Nussbaum, working with other philosophers and social scientists, has identified core human capabilities that are central to living a life with dignity. Any list of central capabilities, as with the list of basic needs, can be tweaked, but such lists are useful in guiding policy formulation. Nussbaum (2004: 47) explains that the ten capabilities she has identified provide an account of the minimal demands of social justice: "a society that does not guarantee these to all its citizens, at some appropriate threshold level, falls short of being a fully just society," she states.

Meeting human needs and providing core opportunities. Maslow's list of human needs, presented only in broad categories, is below and is followed by Nussbaum's list of core capabilities. These are not presented in a hierarchical order, although they are usually presented that way.

1. individual *physiological* needs, related to survival and reproduction (needs shared with other animals in some ways but shaped by human understandings)
2. *socio-physical* needs: the desire for safety, health, bodily integrity, and security (needs that are met by others or in conjunction with others)
3. the need for *self-actualization*: a drive that spurs the creative urge to discovery, self-expression, independent values, play
4. relationship/social-relational needs: the need for acceptance, recognition, love and fulfilling relationships. This involves commitments and shared values.

This list of needs is meant to apply to all societies, but over time what is required to meet these needs changes. Today, for example, one must generally purchase many needed goods and services; this means that access to a job that provides income is necessary for most households.

I have loosely grouped Nussbaum's list of 10 basic human capabilities into four areas below. Her description of each stresses the role that other persons play in determining an individual's ability to exercise each capability.

1. Life, bodily health, bodily integrity
2. Control over one's environment (material and political)
3. Senses/imagination/thought, practical reason, play
4. Emotions, affiliation, including with other species.

These central capabilities cover the needs listed by Maslow (1972), but Nussbaum adds specificity and clarity regarding how society impacts people's ability to function in each of these areas. "Control over one's environment," for example, covers a lot of ground, both political and material. As she puts it (Nussbaum 2004: 49), politically a desirable level of control involves "being able to participate effectively in political choices that govern one's life; having the right of political participation, protections of free speech and association." Materially, it involves "being able to hold property and have property rights on an equal basis with others, having a right to seek employment on an equal basis with others; having the freedom from unwarranted search and seizure. In work, being able to work as a human being, exercising practical reason, and entering into meaningful relationships of mutual recognition with other workers."[7] Each of the ten capabilities is similarly packed with meaning and, she adds, the understanding of many of them has gradually been modified over the years.

Still, the 10 capabilities provide a starting point for societies to consider when deciding on fundamental political principles or constitutional guarantees. The U.N.'s Universal Declaration of Human Rights is one of many documents that can be measured against these points. The Declaration does, in fact, address many of the ten. Rights that are named include the right to a basic standard of living, to medical care, to freedom of expression, to a fair trial, and to take part in the government of the country. The Declaration provides important specifics in some areas. For example, with regard to employment, Article 23 states[8]:

1. Everyone has the right to work, to free choice of employment, to just and favorable conditions of work, and to protection against unemployment.
2. Everyone, without any discrimination, has the right to equal pay for equal work.
3. Everyone who works has the right to just and favorable remuneration ensuring for himself and his family an existence worthy of human dignity, and supplemented, if necessary, by other means of social protection.
4. Everyone has the right to form and to join trade unions for the protection of his interests.

The fact that the U.N. General Assembly adopted this Universal Declaration in 1948 shows that there has long been support for meeting basic human needs. Today there would be additional provisions. The Declaration does not mention LGBTQ+ rights, rights of the disabled or protection of the environment. It does not address specifics of health care, such as reproductive rights, nor specifics of what constitutes favorable work conditions. Even so, the rights set forth over 70 years ago remain aspirational in most signatory countries, although it has had, and continue to have, significant impact globally, including when constitutions are set up or re-examined or when a nation's guiding principles are re-assessed. (It is worth noting that all independent nations of the world are members of the U.N. except for the Holy See in Rome.) In 2015, the U.N. General Assembly adopted 17 Sustainability Goals which incorporate goals inspired by the Declaration, but add and modify the goals in light of the urgency of living on this planet sustainably. These 17 goals are:

1. No Poverty
2. Zero Hunger
3. Good Health and Well-being
4. Quality Education
5. Gender Equality
6. Clean Water and Sanitation
7. Affordable and Clean Energy
8. Decent Work and Economic Growth
9. Industry, Innovation, and Infrastructure

10. Reducing Inequality
11. Sustainable Cities and Communities
12. Responsible Consumption and Production
13. Climate Action
14. Life Below Water
15. Life On Land
16. Peace, Justice, and Strong Institutions
17. Partnerships for the Goals.

Pace of progress is linked to political will. Why has progress toward meeting the goals of the Universal Declaration and the 17 Sustainability Goals been so slow in so many countries, and why would things be different under a non-capitalist solidarity society? The answer lies in the strength of the political will to implement change, which varies by country and over time.

The track record of countries around the globe varies widely, but overall, nations of the world have not lived up to the ideals of the Universal Declaration: poverty, lack of medical care, and discrimination continue. Still less have they reached the 17 sustainability goals. In a few nations, the basic material needs of citizens, if not of all residents, are largely met, but discrimination and inequality persist everywhere, and protection of the environment and sustainability are a universal challenge. The annual report, *How's Life,* published by the Organization for Economic Cooperation and Development (OECD) has data on 11 subject areas for all its member countries, which include the U.S. The report makes clear that every nation has room to improve.

How's Life provides data on expected indicators, such as income, income inequality, years of education, and life expectancy, but also includes measures of air quality, social connections, civic engagement, safety/homicides, social connections, and satisfaction with personal relationships. Data on capital stocks—natural, human, and man-made—are also provided, on the grounds that they are indicative of future well-being. Data are often broken down by sex. The broad range of data available makes it possible to keep track of progress in several areas. These data can also be used to assess priorities.

As this list of subject areas shows, international organizations have moved beyond an earlier narrow focus on Gross Domestic Product (GDP) and per capita income when comparing well-being across countries. Individual countries also are more sophisticated in estimating the well-being of residents. They can compile statistics and survey results from many sources to measure their progress toward the specific goals they have chosen; there are hundreds of variables available that can be used to set up a unique "data dashboard," as Stiglitz (2020) explains.

With so much data and analysis to draw on, why the slow progress that has resulted in today's worrisome state of affairs? The answer to this question lies in the priorities chosen as reflected in the ways in which countries work to reach the U.N. goals. The favored approach can be described briefly as follows: current outcomes in targeted areas are assessed using relevant quantitative variables; policies intended to improve those numbers are then crafted and legislation passed. The proposed solutions often involve improving government services.

But these proposals seldom confront injustices in priorities. They do not ensure that the institutions expected to implement the new policies have the will and the resources needed for success or the capacity to take on expanded roles. Broad political support is critical to adequate funding for programs. It is unlikely that politicians will see the data on an injustice, a proposed solution that requires a realignment of budget priorities, and exclaim: "Why didn't I see that before? We'll get right on it—we'll add a new tax bracket on high-income groups and use the added funds to finance that proposal." If things worked that way, tax rates in the U.S. would have been more progressive ages ago, and needed funding would have been available for health care, early childhood education, and more. Progress, then, depends on the political will to dedicate the necessary resources and effort to achieve the goals set. And that is why a society with deeper democracy and solidarity would do better in advancing the U.N. goals: the political will would be there.

Institutions and Thriving

The political will to achieve a given goal is reflected not only in the level of funding provided, but also in efforts undertaken to ensure that the institutions intended to do the needed work are capable of doing so. This may mean that a shake-up is needed. But what kind of institutional change will result in more effective efforts, and how can the internal functioning of an institution be evaluated, apart from looking at program outcomes after the fact?

Institutions at bottom are social relationships that follow accepted practices. Income, life expectancy, and other quantifiable measures of well-being are the result of relationships in school, at work, and in broader society. To ensure success in meeting persistent problems like poverty or pollution, ***policies must aim at changing institutions, rather than at changing measured outcomes.*** Outcomes will surely fall short if programs are designed and implemented following faulty decision-making processes. Effective institutions, on the other hand, mean effective approaches to achieving desired outcomes.

The institutional goals of government are set in line with political priorities of those in power. This is a basic reason why democracy is important. The services, policies, laws, and programs of government are determined via a political process in which people have a voice. But, as we saw in Chapters 3 and 4, there is a gap between what most Americans want and how their elected representatives vote in Congress. The quality of U.S. democracy has been in decline for some time and political change must precede transformative policy change.

As countries become willing to examine their institutional practices, there are some data related to power and decision-making available. The percentage of women and minorities in law-making bodies, voter opinion polls results compared to lawmakers' policies and similar numbers can indicate problem areas. There are also employment-related data available, including data on the number of discrimination cases tried and outcomes. Such variables can be useful in determining whether people are thriving within an institution. Assessing how well institutions serve society, then, requires use of different kinds of data and different methodologies than those used in, say, estimating future poverty rates. But data related to social relationships within institutions now are only be partial and suggestive. In addition, what data are available are not often considered when making recommendations regarding goals such as reducing inequality.

Assessing the decision-making processes of firms and other organizations yields controversial results. There is little agreement about what power relationships are equitable and little research on the subject. Most social scientists recognize that decisions within firms and government are dominated by those in authority who have veto power over the proposals that come to their desks. The question of whether greater voice for employees is needed or desirable is not often addressed; whether or not greater democratic input from stakeholders would improve outcomes similarly is not much studied.

The lack of equal opportunity, which is related to a lack of voice, is a major reason why some people do not thrive. This is a problem within government itself as well as in most institutions. There is no easy solution. Ending racism, sexism, ableism and other forms of discrimination involves asking probing questions, listening to marginalized voices, and being open to new ways of doing things.

At least one point is clear: these necessary discussions can only take place when the institutions in question are willing to have them. Government and business leaders usually focus on daily tasks within the power structures they've inherited. If they do host discussions about racism, sexism, or worker voice, they will largely control the process. They will be in charge of the implementation of any suggestions that emerge and will mold the recommendations to "fit in" with on-going programs. In brief, suggested changes will be managed from above. Changes that are granted from above can easily be taken away. Deeper democracy, with accountable officials, then, would make a difference and in fact is central to the kind of debates that can lead to lasting institutional change. How power is structured within social institutions determines the policies and practices of those institutions: at the workplace, in the home, and in government. We discuss the politics of arriving at radical/deeper democracy in our leading institutions in a bit.

The Path to a Thriving Society

There has long been agreement that people deserve an opportunity to succeed in life and to thrive. There has also been broad agreement that equity, democracy, and justice are values that should guide society. But the political will to take the difficult steps that would effectively tackle unequal opportunity and injustices has been lacking. The problem lies, not so much with individual politicians, but in the balance of power between capitalist elites and the rest of us, and in all the ways that institutions uphold the prevailing balance of power. From time to time, there are political shifts that change this balance, and progressive legislation is passed. But budget cuts may follow later, eviscerating the intended effect of such legislation.

As noted earlier, deep/radical democracy is a stepping stone on the way to a solidarity society. Under radical democracy, capitalists would be in a defensive position, and progressive agendas could make headway. At the next stage, that of a non-capitalist solidarity society, wealth generated by collective effort would be collectively controlled, and decisions regarding public policy would be made through a deeply democratic process that is not dominated by a non-representative elite. Since such a solidarity society is a ways off, we look at possibilities under radical democracy before turning to imagine the more transformative changes that would be attempted under a solidarity society.

Policies Bringing the Vision to Life

All beings and the planet could thrive once societies dare to care, and share. As we move beyond the minimal requirements for a shared prosperity, we get to the question of what policies might best enable human and planetary flourishing. Or, put another way, if we know the values that we want to guide a solidarity society, how can we begin to translate those values into practice within social institutions today? Values widely mentioned as central to a solidarity society fall into four categories: equity/fairness/diversity; democracy/freedom; community/solidarity/caring; and sustainability/thriving ecosystems. Society has several options for operationalizing these values in practice. Some options could be implemented in the shorter-term under radical democracy, while others would require the longer time horizon needed to arrive at a non-capitalist society. The goals desired, in any case, can only be reached through democratic, participatory efforts backed by mutual support. A democratic, egalitarian society cannot emerge from a process that is dictatorial, dogmatic, or discriminatory. The policies mentioned below can be, and at times have been, implemented in various ways. Note that those policies fostering widespread engagement and empowerment are of especial interest.

Learning from history and from others. There is no shortage of ideas about steps, large and small, that can be taken in the short term to improve everyone's access to the goods, services, and opportunities that enrich lives.[9] In many cases, legislation has already been proposed that would greatly improve healthcare coverage in the U.S. or sustainability in Sweden, and so on. Of especial importance are changes that would incorporate citizens into the decision-making processes that set policy. Changes that institutionalize deeper democracy play a critical role in establishing lasting change.

Once the political landscape has shifted to radical democracy and community-building can begin in earnest, where should efforts be focused? The answer will vary by country and region, in line with the crises being faced. However, economic security, community harmony, and sustainability will all be on the table. The work of departments dedicated to infrastructure, defense, criminal justice, education, agriculture, housing, treasury, and natural resources will need to be re-evaluated. And always, the voices of all whose lives will be greatly impacted by proposed policies should be heard.

A starting point when it comes to crafting a reform agenda for the justice or other system is to examine what that system does, and doesn't do, well. Institutions in different countries, and different programs within those countries, offer models to learn from. Michael Moore's 2015 film, *Where to Invade Next?*, presents examples of practices in countries including France, Italy, Portugal, and more, that have had exceptional success in areas ranging from worker leave policies to school lunch programs, from policing and prisons to education and more. Surveys of some of the possibilities are readily available: the U.N. and the OECD offer detailed descriptions of how various programs work in their member countries, as well as some studies of their effectiveness.

The possibilities are mind-opening. Take the criminal justice system in the U.S. Several programs in both the U.S. and abroad have been shown to reduce drug addiction, recidivism rates, and more. If low crime rates and low recidivism rates are a key goal of the justice system, there is clearly something to be learned from Norway and other countries having success in these areas. One shared value guiding programs for prisoners in countries like Norway is a belief in the possibility of rehabilitation of the incarcerated.

Or consider different electoral systems. Countries with greater participation in elections and policy debates offer a wide range of possible innovations in the political sphere—institutions such as proportional representation, mandated time off at election time, publicly funded policy and campaign debates, and more encourage citizen participation.[10]

Here is a list of some policies and practices that would help people and planet flourish under radical democracy. Most of these would be immediately feasible, if it weren't for political opposition. The policies fall into five broad areas:

1. The basics. Related to income, health, employment, housing, education, and safety.
2. Increased democratic political power. Electoral and campaign reform.
3. A more democratic economy. Revamping of laws regarding business rights. Greater planning.
4. Greater care for the earth. Ending climate change and protecting biodiversity and ecosystems.
5. Stronger, more caring communities. Richer lives, truer justice.

These are areas of concern that are addressed in the platforms of progressive political caucuses and parties, like the Bernie Sanders caucus in the U.S. Democratic Party or the Labour Party in the U.K. under Jeremy Corbyn. Manifestos also present policy proposals. In the Netherlands, a widely circulated manifesto is influencing economic policy debates and we later consider its recommendations. Large movement groups also take policy positions: a variety of policies have been proposed that aim at meeting material needs and at improving access to education, health care, and critical government services. Proposals address deepening democracy as well, including through greater worker voice on the job and greater immigrant rights. Some of these proposals are discussed below.

1. ***Ensuring the basics***

Many countries are ahead of the U.S. in ensuring citizens greater economic security and stronger safety nets. OECD reports provide considerable detail on this. We have already mentioned several programs that the U.S. lacks, but that other countries have. These include: free or low-cost access to education at all levels, universal access to child care and elder care, generous leave and vacation policies for workers, and universal access to healthcare, including reproductive health. Recent thinking in the U.S. and abroad about income security has led to serious consideration of a guaranteed basic income for all. A similar proposal would guarantee a living wage for all employed workers, as well as a guaranteed minimum income for those who are unemployed or unable to work. Free or low-cost education and an end to student debt are other widely supported measures.

Housing is another challenge. Well-designed publicly funded housing for mixed-income groups has met with success in several countries. Various forms of rent control and zoning can also protect the availability of low- to moderate-income housing. In the U.S., community land trusts are a promising innovation. These land trusts allow private homes to be built on community-owned land; the land can't be sold and, when owners decide to sell their homes, the sale price is regulated by the trust.

Several changes related to employment and work are possible. Income security could be provided through guaranteed access to a basic income, but more generous leave policies and better oversight of working conditions are also needed. An expanded supply of higher-quality services in child care and education is important to improving job opportunities.

The value of caring labor should be reflected in improved compensation and economic security for those doing this work. Technological progress reducing work hours should be welcomed, not feared—this only requires taking steps to maintain incomes through shorter workweeks, job sharing, and the like. Job-sharing helped workers in Europe weather the pandemic. Assured upward mobility for those at the bottom of the occupational ladder is needed. Better regulations, with enforcement, are required to protect people in lower-status occupations.

Special attention must be paid to women and minority workers (who often make up the majority of an industry's workforce). Pay gaps need to be addressed. This gap will not be closed simply by ending overt discrimination at the initial point of hiring. In the case of the male–female wage gap, the need to care for dependents is a major factor disproportionately affecting women. Low-income women workers caring for children may want to enroll in training programs, but cannot afford the associated childcare, transportation, and housing costs. Once hired, the initial jump in household expenses caused by these increased costs can be a formidable challenge that should be taken into account by programs creating jobs, like the Green New Deal.

Many low-income workers face abuse on the job, with little legal recourse. Women wishing to escape violence at home also often have few paths available to immediate economic independence, no matter their skill levels. These are other issues to be addressed.

In the long-term, equality at work will entail an end to un-democratic hierarchies at the workplace and collective ownership of large businesses. See below.

2. ***More political power. Stronger democracies with more citizen input.***

People deserve a voice in decisions that greatly impact their lives. The great strength of democratic governments is that the people's demands influence the actions and policies of government. Transparent and equitable political processes increase a nation's unity and stability. Steps can be taken toward better political representation: electoral systems can be improved, campaign reform can end the influence of big money in politics, and measures can be put into place to increase both accountability of elected officials and the impact of citizen input.

Some simple steps can make a big difference. More mandated time off work near election time and for political events would allow for greater civic engagement, and could open more ways for citizens to become politically informed and involved in policy decisions. Other changes would require a significant revamping of current practices. Steven Hill (2015) provides an overview of some practices in Europe that could deepen democracy in the U.S.

Proportional representation, for example, has many advantages. It ensures that political parties that gain the allegiance of a sizable percentage of voters will have at least some seats in legislative bodies. This allows for greater consideration of differing views. As an example: a party receiving, say, 30% of the vote in a district might gain 30% of seats being contested. Proportional voting in the U.S. could be instituted by combining several voting districts into one. When three or more positions are open in the combined district, the chance that a minority party could gain a seat is increased. Proportional voting has contributed to a diversity of occupations and incomes among members of parliament in some countries.[11] Along with larger voting districts, ranked-choice voting could be instituted, which would enable a ranking of candidates by voters and ensure election results are more in line with voter preferences (see http://fairvote.org/[12]).

Public input into budgetary processes, another possible innovation, already occurs at the local governmental level in several areas. The procedures followed could be modified to work at higher levels of government. The White House or Congress could enable citizens to have an advisory vote on different versions of the federal budget. Another possibility: a budget office or offices could be established to analyze the impact of past or proposed budgets on the environment, women, racial minorities, and other groups. Some countries already have such offices. More polling through official government agencies to determine public preferences regarding proposed legislation could improve the legislative process. Tax reform is called for: economic justice demands that taxes be realigned in accord with the concentrations of income and wealth.

Options for campaign and electoral reforms include: 100% public funding of campaigns or a maximum donation allowed per individual and per group. Free media time for qualifying candidates and parties, streamlining of the voting process through universal/automatic voter registration along with online balloting and other changes would improve voter participation.

Several European countries have a practice called Question Time that the U.S. and other countries could emulate. During Question Time, top government officials from the prime minister down are questioned by members of opposition parties. The sessions may be televised weekly and can provide jolly good entertainment.[13]

3. ***A democratic economy: revamping of business rights. More planning. Stronger regulation.***

Oversight of business. Laws that set business rights and regulate its activities should protect the interests of the public. To ensure firms' compliance with laws and regulations, large firms could be required to obtain a federal charter to operate in the U.S.; penalties would be imposed on firms for repeated violations of regulations. Loss of limited liability could be a possible penalty. Oversight of critical industries should be strengthened, in order to protect both the environment and people's access to electricity, clean water, the internet, phone service, and pharmaceuticals. A first step toward improved oversight would be the collection and dissemination of comparable performance data by state. This would show where government intervention is needed to ensure acceptable performance.

The growth of alternative businesses that have a commitment to employees and communities can be fostered. These businesses are already a large part of the U.S. economy and are also important in other countries. There are roughly 11,000 employee-owned companies in the U.S. alone, and about 130 million people are members of cooperatives of one kind or another (including credit unions and mutual insurance companies). Employee stock ownership plans (ESOPs) cover about 13.5 million employees.[14] There are also nonprofit enterprises, public–private partnerships such as community development banks and corporations, state-owned enterprises, and other non-traditional businesses that are committed to local economic development and social responsibility; they have changed regional growth patterns across the country. The Democracy Collaborative, the Center for a New American Dream, and other progressive organizations are promoting the growth of such mission-minded, alternative businesses. Many are discussed in Schor (2011) and Alperovitz (2013).

Firms with tens of thousands of employees are social enterprises that can rival or eclipse entire nations in size. Their actions create waves that are tsunamis for those caught in their path. Taming these beasts is an urgent priority. Having a few stakeholder representatives on the boards of these firms will not suffice to safeguard public interests. While most should be reduced in size, as long as giants remain, a majority of the board of directors should primarily represent the public interest, as Gar Alperovitz and others have noted.[15]

Stakeholders from shareholders and employees to communities and consumers have rights that can only be met through adequate representation on a firm's board. Allen White of the Tellus Institute (White 2013) proposes transferring 50% of a company's stock to a firm-specific public trust. Perhaps 1 or 2% would be transferred yearly, until the 50% threshold is reached. The trust, in turn, would control 25% of director seats; stakeholders, 25%; other shareholders, 50%—but with shareholder voting rights depending on the duration of their investment. There have been several other proposals for more democratic ownership of firms. In Sweden in 1975, the Meidner proposal would have gradually transferred majority ownership of firms to employees. (The bill as finally passed in parliament, however, was considerably more modest in its scope.[16])

To ensure that collaborative ownership of firms functions well both for the firms and for the public interest, an appropriate process for choosing board members is critical. One way of deciding who joins the board of directors, in addition to employee representatives, is through a nomination process that involves the grassroots. Candidates for training and eventual appointment to boards could be chosen from names put forward by stakeholder organizations that do not depend on corporate funding, such as neighborhood associations, labor unions, and community service clubs. These candidates would be subject to approval by a locally elected body, screened based on candidates' experience, background, and specified criteria, for example, relating to sustainability and anti-discrimination credentials.

The training could be done through universities' business administration schools and financed by a small tax on trades on financial exchanges. Once trained and elected or appointed to a board, these individuals would be paid from a common fund financed by the corporates using their services. The appointees would have access to experts as needed for their work, paid for by the company.[17] Business consultants have long recommended that a certain percentage of board members be professionally trained and independent. This process builds on that recommendation. See appendix for more on these suggestions.

Once nations reach the point where they make such demands of corporates, other similarly radical demands regarding business's social responsibilities will be possible.

Finance and banking. Deregulation of finance, which enabled banks to invest in highly risky assets, led to the crash of 2008. Another similar crisis remains possible, as we saw in Chapter 2. Restrictions on risk-taking post-2008 have scarcely dented the risk exposure of the largest banks and so must be increased until they are effective. Banks as they extend credit create money, which affects all sectors of the economy. In this role, banks are meant to channel funds into investments of economic benefit. When banks fund housing, infrastructure, and similar projects, they aid community development. They are providing a public service, and there is an argument to be made that banks should be public utilities. In a society in which banks served the public good, economic development would be a priority goal and loans to projects that are detrimental to society would not be granted. Loans to companies that routinely break federal regulations would also be denied.

The Federal Reserve Board should be restructured to ensure greater independence from the banking industry that it oversees. The debates around the creation of the consumer protection agency revealed some of the shortcomings of the current system. Financialization of the economy must be reversed. The size of the sector and the share of profits it receives should be more in line with its value to society.[18]

More planning is needed to smooth the transition to sustainability and to more equitable workplaces. Localities adversely affected by shifts in economic activity should be involved in the formulation of programs to address the loss of employment and tax revenues. Participatory planning can expedite the achievement of shared goals (se Hahnel 2012).

The division of labor is a political as well as an economic question. Oversight and intervention are required to ensure an equitable division of labor and equitable treatment of workers. One key issue is caring labor. Most of us will be both care givers and care recipients in the course of our lives. We have a shared responsibility as a nation to ensure that care given is adequate and that care givers are adequately reimbursed. Currently, care is under-valued, and under-shared. The Care Manifesto (Care Collective 2020) provides insight into how to begin the necessary adjustments.

No one should be relegated exclusively to drudge work. Career ladders opening opportunities should be available to all, and all jobs, regardless of skill levels required, should pay a living wage. Greater sharing of less-favored and more-favored tasks can be facilitated.

Public service programs could provide an alternative to the military for young people. An obligatory year of public service after high school should be considered. It could benefit both the nation and those participating.

Revised federal-state-local relationships. As the population of the U.S. has grown, the number of persons and the variety of issues faced means that the federal structure needs to become more flexible, with more authority delegated to the regional and/or state levels. Constitutional changes may be needed for regional associations to get appropriate decision-making power, e.g., in regard to energy.[19]

4. *Care of the earth*

A shift toward sustainability in all areas of activity is basic to stopping climate change: greenhouse gas emissions must drop to net zero by 2050 if the internationally accepted goal of limiting the increase in the earth's atmospheric temperature to 2 °C is to be met and catastrophic consequences averted. The 2022 report of the U.N.'s Intergovernmental Panel on Climate Change (IPCC) makes it clear that today's socioeconomic (capitalist) system is unsustainable and must give way to a transformation of social institutions and practices.[20] The disruptive changes required to move from fossil fuels to clean energy sources will only be politically and socially feasible if national resources are redistributed to bring net benefits to low- and middle-income people during the transition. The slowing growth rate of the world's population, which had dropped to 1.02% in 2020, makes this a more manageable task. [Growth rates have fallen in several countries with formerly high growth rates, including India (1.0%) and Brazil (0.7%).][21] The IPCC also speaks of "Integrated, inclusive planning and investment in everyday decision-making...(to) increase adaptive capacity...." It adds that "equitable outcomes contribute to multiple benefits for health and well-being and ecosystem services, including for Indigenous Peoples, marginalized and vulnerable communities" (IPCC 2022 Report Summary: 32).

Protection of the earth's ecosystems and biodiversity involves addressing issues from habitat protection and species extinction to plastics pollution and toxic chemicals production. Agricultural practices need to become sustainable and regenerative, absorbing more carbon, enriching the soil, and enhancing food quality. Major portions of land and sea must be conserved.

International organizations from the IPCC to the World Wide Fund for Nature, as well as national organizations, both public and private, and local groups all have compiled recommendations for protecting the atmosphere and the environment. Businesses around the world are developing and selling new technologies that better protect the planet. The work of the IPCC is particularly worthy of note. It is comprised of 195 member countries whose experts assess the scientific research related to climate change and issue shared conclusions. Their work informed the goals set by the 2015 Paris Agreement and its subsequent updates. Working Group III of the IPPC examines the options available for reducing greenhouse gas emissions and issues periodic reports.[22] [Greenhouse gases, although they account for no more than 0.1% of the atmosphere, are responsible for the earth's having an average temperature above zero degrees Fahrenheit. The average is closer to 57 degrees.[23]]

The U.S. Green New Deal (incorporated into bills introduced under the Biden administration) has a series of provisions aimed at building a strong green infrastructure, transitioning to renewable clean energy, low-emission transportation systems, and low-emission heating for buildings. Other countries are ahead of the U.S. in their implementation of green policies. Some recommendations aim at more drastic and immediate change: these include terminating the expansion of air, sea, and road travel; promoting local food supplies; and ending the use of pesticides and fertilizers in agriculture. Promising research addressing climate change is being carried out worldwide.

Some recommendations bring a smile. One surprising way to cut methane emissions is by adding a small amount of seaweed to cows' diets: a study found this reduced their methane emissions by 90% [cattle are a significant source of methane].[24] Another unusual step, featured on *60 Minutes*, is restoring reindeer, bison, and other grazing herds to the Arctic region, which can prevent up to 80% of permafrost from thawing.[25]

More recommendations are low-cost and highly effective. Cement production accounts for about 8% of global carbon emissions originating from human activity. If cement production were a country, it would be the world's third-largest emitter. Researchers at MIT are among scientists who have developed low-cost cement production processes with zero CO2 emissions. Sweden in fact is about to begin production at one such zero-emission plant.[26]

Unfortunately, the drastic cuts to carbon emissions needed during the coming decade are unlikely to happen under our current system. The dire consequences of rising global temperatures are made clear in the IPCC reports and in other studies, including reports issued by insurance companies and the U.S. Department of Defense. The Defense Department is not focused on falling crop yields, fires, weather events, disruptions in supply chains, massive climate migrations, warming oceans, and sea-level rise per se, but rather on the consequences of these events for increased national hostilities.[27]

Grassroots efforts at every level will be critical to ending and reversing climate change. Activists have succeeded in bringing widespread change. Recycling, efficiency ratings, and solar panels are now common. States and universities are starting to divest from fossil fuels. Thousands of initiatives are making a difference, yet so far these have not been enough. The most comprehensive listing that I have found of needed actions, along with leading organizations and individuals tackling the issues, is at: https://regeneration.org/. Paul Hawken (2021) is lead author of the associated book.

5. ***Stronger, more caring communities. Richer lives, truer justice.***

Communities are built on shared experiences. Sports events, parades, and fireworks, for example, strengthen community bonds. Everyone should have access to such events, but also to the arts and cultural happenings. Everyone should be able to participate in recreational activities and in local municipal policy debates, unhindered by a lack of income or lack of transportation. Civic participation can be encouraged by regularly scheduled public debates on local and national issues. Mandated days off work can facilitate attendance.

Changed employment practices also need to be part of the effort to strengthen communities. Working families require sufficient leisure time to recreate and participate in the life of the community. Communities with stable networks are better able to react with resilience to new challenges and better able to innovate solutions to problems faced. Large firms too often engage in employment practices that are disruptive to communities, moving employees around the country rather than investing in arrangements that would work better for employees and communities. Unemployment benefits need to be sufficient in value and duration to allow workers to stay where they are for a reasonable period of time, and every worker, including gig workers and part-time employees, should be covered by unemployment programs and health insurance. Immigrant workers in this country are caught up in on-going struggles against often egregious injustices; these wrongs, which are widely documented, require redress.[28]

Part and parcel of changing dysfunctional social behaviors is acknowledging and addressing them. Reform of the justice and prison systems would be part of this. Belief in the possibility of rehabilitation motivates many successful programs around the world. At the local level, truth and reconciliation commissions or similar can be valuable in redressing wrongs perpetrated against Indigenous peoples, Blacks, and other minorities. Nationally, also, much remains to be done. Restorative justice programs can help heal harms to people and community whether or not those harms have been addressed by the criminal justice system. A justice system that is more compassionate and just would be set within systems of services that provide mental health and addiction treatments and services that enable successful rehabilitation and employment. This was a major reason behind the demand to channel funds away from the police and to social services issued by the Black Lives Matter movement.

Discrimination is a dysfunctional behavior that divides and harms communities. People discriminate on the basis of characteristics such as race, sex, gender identity, religion, immigration status, disability, and more. Ending discrimination depends not only on legislation but also on soul-searching and communication; it requires that those who discriminate understand why their thinking needs to change and also that they commit to expending the effort needed to do so. Insight into the injustice of discrimination can be imparted not only through study, but also through dramas, storytelling, poetry, and moderated conversations. Educational institutions have an important role to play. Governments at all levels should fund efforts to ensure that members of disadvantaged groups are treated equitably. Restitution is called for as a means of repairing some of the wrongs of slavery and of violence against minorities, including the violation of the rights of Indigenous peoples. Research is needed that documents and presents the case for fairer treatment of disadvantaged groups, using data, case histories, and more. The disadvantaged groups themselves should have oversight of this research, through their organizations.

Julie Matthaei (2017) and others in the U.S. Solidarity Network have identified several solidarity processes that are being used to address and dismantle inequalities, and to build qualitatively different relationships.[29] Equal opportunity struggles, one of these processes, works to end domination/subordination dynamics and the related stratifications and polarizations. The integrative process brings together the many processes challenging social divisions of all kinds, as well as the perceived human division from nature. There are a myriad of ways in which these solidarity processes are being put into practice, from time trade circles and community gardens to cooperatives and local currencies.

"Schools of hope" (a term used by Martha Nussbaum [2004]) are important in building community. Belief in the possibility of a better future is critical to making it a reality. "Schools" mentioned by Nussbaum include the arts, critical dialogue, protest movements, the study of theories of justice, religion, nonprofits, and national service programs.

International policies including those regarding immigration policies must respect human dignity and human rights. Support for dictators and repressive regimes should be halted.

Whichever of the above policies are being pursued, there is a need for processes that allow for democratic input. This list gives some indication of the many areas in which creative thinkers have offered suggestions that, if properly implemented, can be expected to address the problems targeted with positive effect. But we are not yet at the point where radical proposals are close to national implementation; in addition, most have not been fleshed out. As Gar Alperovitz states:

> Much rigorous investigative research into essential questions—around democratized economic institutions, participatory mechanisms for planning, assessments of scale for effective governance—is simply not available. Gar Alperovitz—*Principles of a Pluralist Commonwealth.*[30]

Getting There from Here

Rationalizations abound when it comes to explaining why ending homelessness or hunger isn't a priority in higher-income countries. There is much political work to be done before elected leaders show a commitment to taking the difficult steps required to redistribute the tax burden and fund the initiatives that would bring greater equity, sustainability, and deeper democracy. Political priorities must change if we are to thrive.

Our experiences in crises—including the pandemic, the financial crisis, and Hurricane Katrina—have shown us that the ruling elite routinely fails to take even the indispensable, urgent steps needed to keep citizens alive and well, let alone steps needed to enable them to thrive. A new socioeconomic structure, with new values and goals—a radical democracy—would be able to lead a collaborative effort that could bring the goal of achieving a thriving citizenry and planet, a solidarity society, into the realm of possibility.

Under radical democracy, capitalist plutocrats would have a minority voice, unable to control political agendas. This has happened before in Europe, as we have seen. Some of the possibilities opened by such a development are indicated by the policy options just discussed.

But more could be accomplished if capitalism should give way to a post-capitalist solidarity society. The dictum of Karl Marx—From each according to her ability, to each according to her need—offers one ideal. This ideal would at the least influence thinking in a solidarity society. In any case, a post-capitalist society is one in which there is collective ownership of the major means of production, including public controls over land. This ownership could take various forms: cooperatives; public ownership at the local, regional, state, or national level; ownership by a public trust with management oversight shared by employees, communities, and/or elected officials; control of companies could be vested in a board with the majority of seats held by representatives of stakeholders; and so on.

There would be planning at local and national levels to enable the shift to sustainability, to ensure affordable eco-friendly housing and transportation, and to end periodic recessions. Some idealists even see an eventual end to wage labor: everyone could count on a decent standard of living regardless of the work they did, whether at home or elsewhere. The division of labor into specialized areas would be reduced as possible, so that no one would be relegated exclusively to drudge work; there would be greater sharing of less-favored and more-favored tasks. Caring for one another would be recognized as a responsibility that is collective as well as individual. Who would do caring labor, whether paid or unpaid, and how they would be reimbursed, would be acknowledged as a political policy question to be decided democratically and continually adjusted, as argued by Tronto (2013).

Michael Yates (2018) describes this imagined post-capitalist, radically egalitarian society at some length. As Yates puts it, there would be:

"equality in all spheres of life—between men and women, among all racial and ethnic groups, among all people irrespective of their gender identity or sexual preference, among and within every country with respect to work, region, and access to all social services and amenities."[31] There would be an end to inequalities in power, except as delegated democratically, and no hierarchies (such as patriarchy) extending into most of life. There would be an end to racism, imperialism, and plunder of the natural world. This non-capitalist egalitarian society would embrace "whatever is social, collective, sharing, and unalienating."

This thriving, caring community is a long-term goal and may never be attained. But it is certainly possible for a country to establish a society with goals including those described by Yates.

In the short term, efforts must focus on creating a radical democracy under capitalism, but the long-term goal of a thriving society should influence strategy today. Immediate demands will focus on urgent issues in sustainability and human rights, but new policies which provide for greater stakeholder voice will also provide lasting leverage for future progress.

Creating the unity needed to advance toward radical democracy and then a solidarity society depends on raising awareness of the need to shift power away from capitalist plutocrats. This work has begun; there is a growing understand of the causal links between inequalities in power and economic inequalities and other social ills (e.g. Shutz 2011, Pickett and Wilkerson 2009). There are several possible approaches to deepening democracy and establishing a solidarity society. Each has positive points and, as Erik Olin Wright (2019) argues, these can be combined in an approach he terms "eroding capitalism."

Eroding Capitalism

Wright identifies six different strategies that are intended to establish a non-capitalist, egalitarian, sustainable, and emancipatory economy.[32] He terms the first five: escaping, resisting, taming, dismantling, and smashing capitalism. The sixth, eroding capitalism, combines elements of the first five; it is the approach he favors. Each of these approaches recognizes that the domination of society by persons driven by the pursuit of wealth means oppression of those doing the paid and unpaid work of society. Each approach recognizes that values other than market values deserve the place of honor in society. Wright groups these values under the triad of equality/fairness (including to future generations), democracy/freedom, and community/solidarity.

After a brief discussion of the various ways in which people have rebelled against the dehumanizing practices of capitalism, Wright notes that eroding capitalism depends on resisting and escaping capitalism through alternative ways of being and doing as much as it depends on the concentrated politics of taming and dismantling it. He explains that the Amish, as they attempt to escape the capitalist way of life, contribute to our belief that collectively we can choose a different path. Artists as well spark our imagination, opening us to new possibilities and new ways of resistance.

Wright argues for democratizing economic life and the state. The voluntary, cooperative exercise of power is the only path leading to its realignment. There is transformative work to be done throughout society. Alternative forms of business activity can change the incentives driving production for the market. New forms of citizen participation in politics and new institutions for democratic representation can invigorate political activity. Organizing efforts at the grassroots can create new political formations capable of overcoming the challenges of privatized lives and competing sources of identity.

All the forms of resistance to the oppressions of capitalism listed by Wright are on-going. There are communities from the Amish to farming collectives guided by motives that are in service to life-giving values. There are organizations and businesses devoted to service rather than profit; they are resisting the lure of the capitalist drive for ever-greater wealth. (In the U.S. alone, there are over 1.5 million tax-exempt nonprofits under the 501c designation. And, in addition, most employment is not in the very large firms, but in smaller firms/organizations or in the public sector.) There are many politically involved groups working for legislation that would "tame" capitalism through reforms in all major areas—the criminal justice system, health care, environmental protection, and more. And there are political parties or party factions with broad agendas aimed at making this a more just and sustainable society, and some groups that want to end the privileges and protections of the wealthy under capitalism by, in effect, "smashing" it.

Let's quickly look at the unacceptable "smashing" option. In the past, revolutions that aimed to establish a workers' state have failed to build an egalitarian, democratic, and caring society in the aftermath of their ascent to power. Seizing power through either insurrection or elections does not bring an immediate change in the institutions and ideologies that permeate a country. Rupture with the old system will not automatically result in a path leading to fulfillment of the revolutionary goals. A long process is required to change the values and thinking that permeates social institutions.

Are we moving toward transformative change? If all the collectivities working for a better world were on the same page, progress would be rapid. Political theorists and scholars looking at the question of how to build unity among the many different agents of change have examined their goals and values, as well as the internal practices of their organizations. The key to unity, it seems, is shared goals springing from a common adversary, mutual respect and a sense of collective identity.

Thinkers in the critical theory tradition have laid out the promise and pitfalls of building coalitions for progressive change today.[33] While they acknowledge the contributions being made by many organizations, they often note that political parties will eventually have an essential role. Coalitions that do not enter the national political fray (vertical engagement) are limiting their potential impact.[34] Only political parties are able to shepherd broad agendas for change through a legislative process and into mainstream practices. Other collectivities can shape the values and goals of political parties and, through such horizontal organizing, provide the grassroots mobilizations that give a party power. But it is not wise to believe that uncoordinated progressive efforts will suffice to bring about transformed political institutions. Societies do not automatically evolve toward the greater political good; coordinated political effort is essential.

Left Populism: The First Step Toward Radical Democracy

In today's unsettled and uncertain times, the middle ground is not the safe haven for politicians that it once was. Rather, at this juncture, there is a political opening for populism on both the right and left. As we saw in an earlier chapter, Mouffe and other political theorists believe that the consolidation of left populism is the political imperative of our time for those working toward radical/deep democracy. Left populism unifies groups around shared values and broad themes; it can become a powerful political force leading to a deeper democracy. We briefly look at recent political developments and analyses that examine why groups on the left have failed to offer an adequate response to the growth of the right wing in the U.S. and other countries, and also offer some conclusions about the road ahead.

The rise of Trump and the growth of right-wing extremists here, as elsewhere, is linked to the fact that people do not identify with mainstream leaders who do not recognize their unmet needs/radical demands. They are looking for political leaders who articulate their frustrations and hopes. Progressive politicians in Europe, in countries as diverse as the U.K., Spain, the U.S., and Greece, have recognized the reasons for this discontent and have been able to win broad support for progressive party platforms, built on values of justice and equality. This growth of left populist movements bodes well for the forging of a collective will to establish deeper democracy.

Recall that radical democracy keeps and strengthens democratic institutions, including multi-party politics, a free press, universal suffrage, trial by jury, and all the basic institutions revered by democracies today. Radical democracy is a deeper democracy in which political debate is not dominated by the concerns of wealthy plutocrats, and is conducted in an agonistic, not antagonist, manner. That is, there is conflict but opponents treat each other with respect and not as enemies. This radical democracy could be built once the political parties in power are ones dedicated to addressing popular progressive demands, such as those for economic security, inclusive politics, environmental protection, and assured access to high-quality services including higher education and health care.

A taste of what radical democracy would be like comes from countries where strongly democratic Labor or Socialist parties have risen to power, such as in Sweden or Norway (see e.g. Lakey 2017). These countries have at times had parliaments that were dominated by political representatives who were elected due to strong popular backing and who had little obligation to the wealthy. The leaders of these democratic socialist or labor parties were then in a position to establish strong social safety nets, and the wealthy few had little choice but to agree to legislation favorable to the many.

The transformation of politics under radical democracy does not bring an end to capitalism, but is a step in that direction. It is the result neither of reforms put in place through the usual channels nor of a revolution that occurs outside of democratic processes. Its emergence depends on a new, progressive party or coalition of parties coming into power.

The transformation of state hegemony under radical democracy will not develop smoothly from the gradual spread of good things like cooperatives or progressive online networking. Capitalist oligarchs, after all, will mount strong opposition to any effort to end their hegemony. Rather, the growth of a left populism that unites the major social movements and their allies in political parties is the necessary precursor to the political battles leading to radical democracy.

Capitalism will not disappear in one fell swoop. But radical democracy in the U.S. would likely mean a turn toward social programs common in Scandinavia and Europe, and toward other social, political, and environmental practices aimed at advancing sustainability and equity. Which reform initiatives rise to top priority will be decided organically within the parties. The agendas of left populist parties in recent years have varied by country, but have included nationalization of key sectors, increased governmental oversight of business, and greater employee representation on business boards. And also part of these agendas have been policy changes that address equality, sustainability, social welfare, stronger communities, and increased democracy.

Chantal Mouffe, in *For a Left Populism* (2018), discusses why the left has failed to win strong populist support for its progressive agendas.[35]

Us versus them Perhaps the most serious shortcoming of left politics lies in its underestimation of the importance of affect in the creation of political identity. The emotional bonds developed within a group, Mouffe explains, are basic to that group's collective sense of identity and to political commitment and action. Only a collective "we" can work as a unit for the goal of displacing a collective "them."

In politics, the creation of this collective "we" rests not only on shared values but also on shared experiences and importantly on a united opposition to an adversary. In this case, the adversary is the wealthy elite who dominate the country. A shared recognition of the adversary is necessary to unite a collective opposition. This stance in an organization can originate from any of several radical, democratic goals. The labor, environmental, feminist, and other social movements each have radical projects that cannot be completed under today's capitalist power structure. The leaders of social movements may not immediately recognize their common need for a transformed state and new governing principles, but over time political experience will make clear the radical nature of their demands and the need to join in coalition to end the plutocracy that prevents the achievement of their goals.

Recent presidential candidates in the U.S. have attempted to gain populist backing, with some success. As a Presidential candidate, Barack Obama was able to build the sense of a collective "we," although his message was not radical: no adversarial "they" was identified. Bernie Sanders, on the other hand, identified oligarchs as the adversary, but was unable to create a sufficiently encompassing "we," as is also necessary. Mouffe makes several other points in her analysis of the left:

- Too often progressives have assumed that economic issues related to work are the primary political concern of working people. The importance of other issues has been underestimated, including such concerns as justice system reform, LGBTQ rights, women's rights, non-discrimination, and sustainability. To succeed, a populist movement must be inclusive of all urgent demands and must not limit itself to a few select demands a priori.
- Some political movements are too narrowly focused on a single demand, failing to link with others in common cause. Support of local food, local business, organic farming, recycling, rent control—all are positive efforts. But if movements do not engage with political parties and mainstream institutions, they will have limited impact on the broader capitalist hegemony. That is, vertical engagement, as well as horizontal, is necessary.
- In particular, political parties are indispensable in the articulation of broad political identities. People locate themselves politically in reaction to those parties. The arts, narratives, and common projects are also integral to the process of building identity. But without collective confrontation of dominant political parties, fragmented efforts at reform will be limited in their effect.
- Some progressive groups underrate the importance of participatory democracy within their organizations and within the governments of the future. Though the current democracy is broken, democracy is vital to freedom, and new channels must be opened. Bottom-up democracy cannot be built through top-down means. I will add that tackling racism, sexism, and other negative ideologies that stratify and divide requires conscious effort by progressive groups, beginning within their own organizations. To fulfill its promise, left populism must be deeply democratic.
- The site of political renaissance will be national and local, not global, although activists will link their work to other nations and take positions on international issues. The personal experiences and discursive practices that bring people into a common political bond necessarily occur largely at the local or national level. This means that no one agenda will fit populist parties in all countries, and the base of these parties will differ as well, affecting their demographics and more.

Policies initially enacted by a radical democratic party might tackle election campaign reform, farming practices, the minimum wage, paid leave, health care, housing programs, or policies aimed at reducing net greenhouse gas emissions. Such measures can be, and in many cases have been, enacted in dominantly capitalist economies, without transformative change in human relationships in productive activities.

So, while radical democracy would be a giant step forward, more work would be needed to achieve non-reversible progress toward a sustainable, solidarity society. A sharp turn to clean energy and an end to the use of fossil fuels—imperative for sustainability—requires the commitment to both sustainability and justice that characterize a solidarity society.

Inclusive processes are integral to the work of building a solidarity society. Battles must be waged on multiple fronts. Resistance to dehumanizing practices—practices with compellingly unjust effects—will be shaped by the specifics of each kind of abuse of power. Consider the power relations of patriarchy versus capitalism, for example. The differences in oppressive practices create different modes of resistance. As MacKinnon (1993: 437) notes succinctly, "sexuality is to feminism what work is to Marxism: that which is most one's own, yet most taken away."[36] Resistance to power based on gender, then, requires developing liberating theories of sexuality, while a new understanding of work is basic to transforming the workplace.

Each society handles internal political conflicts differently. If a society has institutional processes that make space for discourse that is inclusive of both advantaged and disadvantaged groups, the society may be able to avoid social disruptions ranging from armed conflict and politically motivated terrorism to crises in health care or the environment.

Happening Now and Next Steps Toward Solidarity

Millions of people are now working to make the world a place where people can thrive, and many of them would support the policies of a deeper democracy, like those listed above. They are organizing through political parties and caucuses; through coalitions of social movements; and through nonprofit organizations, both those self-funded and those supported by foundations and industry. They work within public agencies and international organizations, and as community volunteers. They are the leaders of social change today. Their creative work has the potential to win broad support and excite the momentum needed for the transformative change the world so badly needs. We discuss some of their efforts below.

Within political parties Several progressive populist leaders have received widespread support in recent years. U.K. Labour Party membership under Jeremy Corbyn grew rapidly from 2015 to 2018, becoming the largest party in the U.K. (although it lost in the 2019 general elections). In Greece, the young left populist Syriza party managed to rise to power in a coalition government in 2015. In Spain, Podemos skyrocketed in numbers between 2014 and 2018, and in the U.S., Bernie Sanders came close to winning the Democratic Party's nomination for the presidency in 2016. At about the same time, the Green New Deal (incorporated into bills introduced in Congress in 2021 under Biden) gained the support of the Progressive Caucus of the U.S. Congress and of hundreds of organizations.

The Labour Party Manifesto of 2015 under leadership of Corbyn marked a radical turn for the Party. It called for eventual state ownership of key firms in the energy sector and for the mail and railway service to be returned to state ownership, with state directorates to oversee key sectors of the economy. The manifesto addressed jobs and the quality of work (both worker and stakeholder rights); education; inequality and discrimination; macroeconomic policies (finance, taxation, industry, green energy, trade); richer lives (arts, recreation, outdoor activities, sports); deeper democracy; and international issues, as well as income levels, health, and physical security.

The appeal of the revamped Party agenda is reflected in the tremendous growth of the party: there were 198,000 members prior to the 2015 election and approximately 552,000 full members as of December 2017, making it the largest political party in Western Europe. The Party lost ground after 2018, largely due to its wavering position on Brexit. Membership was at 496,000 as of November 2020.

In Spain, the left populist party Podemos grew from 100,000 members in 2014, its year of formation, to 500,000 in 2018, becoming a force to be reckoned with. It called for public control of key sectors from energy to pharmaceuticals.

In Greece, Syriza came to power largely on the basis of its opposition to the austerity program being imposed by international lenders. It also called for the nationalization of banks and key utility and service companies. Syriza was established in 2004 and in 2009 legislative elections gained less than 5% of the vote. But in 2015, the party received 36% of the vote, giving it the largest share of seats in parliament. It then formed a coalition government which took power in the country.

In the U.S., Bernie Sanders led a left populist movement that gained surprising strength within the Democratic Party, based largely on his call for a more equitable distribution of wealth. In 2016, he lost a close race with Hillary Clinton for the Democratic nomination to the presidency. Sanders called for breaking up big banks, taking big money out of politics, the promotion of coops, forgiveness of student debt, and higher taxes on the wealthy. The political action organization, Our Revolution, which was spun out of his campaign, regularly supports progressive candidates in their bids for public office. In 2020, another contender for the Democratic nomination, Elizabeth Warren, had a platform similar to that of Sanders. She called for large corporations to be federally chartered and, in the case of billion-dollar firms, 40% of the seats on the board of directors to be elected by employees.

The U.S. Congressional Progressive Caucus is another group often supporting significant changes. In 2020, the Caucus endorsed the THRIVE Agenda, the version of the Green New Deal laid out in HR 1102. This agenda calls for limiting the rise in global temperature to 1.5 °C; it calls for investments in communities of color; and it lays out policies aimed at enabling "greater racial, economic and gender justice; dignified work; healthy communities; and a stable climate."[37] This Green New Deal is backed by hundreds of organizations; much of its agenda was incorporated into Biden's Infrastructure Bill, which passed, and into the Build Back Better bill, which stalled.

Coalitions Flexing their Muscle In addition to this movement toward the left within political parties, globally several coalitions have formed around broad, encompassing demands that call for a redirection of the economy and of political goals. We now look at some of these, beginning with a manifesto impacting political debate in the Netherlands, then at the THRIVE agenda in the U.S., at the manifesto *Feminism for the 99%*, at the demands of the Movement for Black Lives network, at the democratizing work statement signed by thousands of scholars around the world and more. All share a commitment to working for greater equity and sustainability, but not all express concern about capitalist plutocracies.

In the Netherlands, a manifesto signed by about 200 scholars in the spring of 2020 is being widely cited during political debates about the way forward.[38] There are both points of agreement with, and of departure from, the policies set out in the Green New Deal and other progressive agendas. Five points are emphasized in the brief document, *Planning for Post-Corona: Five proposals to craft a radically more sustainable and equal world*:

1. a move away from "development" focused on aggregate GDP growth;
2. an economic framework focused on redistribution;
3. transformation toward regenerative agriculture;
4. reduction of consumption and travel;
5. debt cancellation.

This document is an excellent starting point for discussion of what should be included in party platforms addressing today's crises. Their main points are summarized below; key differences that set these recommendations apart from the THRIVE Agenda are indicated. According to this document, the Netherlands needs to:

1. Move away from "development" focused on aggregate GDP growth. Economic policies should instead focus on end goals for each sector. A redistribution of production is required—growing and improving output of sectors such as clean energy, research, recycling, green public transportation, education, child care, health care, and other critical public services, while de-growing other industries due to their unsustainability or their role in driving excessive consumption (especially fossil fuels and toxic chemicals, but also advertising, plastics, and others). The THRIVE agenda, in contrast, does not directly address the need to downsize some industries apart from those in fossil fuels. Instead, it calls for standards that would protect the environment.
2. Craft an economic framework focused on redistribution, one which establishes a universal basic income rooted in a social policy system that covers everyone; a strong progressive taxation of income, profits, and wealth; reduced working hours and more job sharing; and adequate valuation of care work and essential public services such as health and education. Unlike this recommendation, THRIVE does not mention taxation or reduced working hours.

THRIVE would require fair employment practices in all jobs created under the Green New Deal, including measures to protect workers' right to organize.

3. A transformation toward regenerative agriculture would make agricultural production sustainable and conservative of biodiversity, with fair employment practices including working conditions and compensation. There would be more emphasis on local and vegetarian food production. THRIVE advocates change that would raise the incomes of small farmers and promote regenerative agriculture.
4. A reduction of consumption and travel, with a drastic shift from luxury and wasteful consumption and travel to basic, necessary, sustainable, and satisfying consumption and travel would improve lives. There would be less planned obsolescence and fewer throwaway products; less heating and travel based on fossil fuels along with more consumption, and availability, of educational programs, community-building events and healthy outdoor activities, and the like. THRIVE does not address the need for changes in the composition of consumer expenditures beyond calling for sustainability and for disadvantaged communities to have greater access to vital goods and services.
5. Debt cancellation, especially for workers and small business owners and for countries in the global south (of loans from richer countries and international financial institutions). In the U.S., student debt cancellation is needed. Unequal protection from the burden of debt is a recurring problem during crises and long-term solutions are needed. The THRIVE agenda calls for equitable access to education, including to industry training programs, but does not mention current debt loads.

One point that is implicit in the manifesto as part of redistribution should be made explicit: an end to all forms of second-class status among people is of critical importance. Here THRIVE is the stronger document, with frequent references to the many inequalities and injustices that disadvantage some groups and with recommendations to address these wrongs. Also, the manifesto has no demand for a redistribution of political power while THRIVE concludes with a call for a renewal of public institutions, invigorating them through democratic governance and accountability. Neither the THRIVE agenda nor the Netherlands manifesto address the need for financial reforms and the definancialization of the economy.

Authors of the Netherlands five points want to see their program fleshed out and enacted as part of a post-COVID recovery package. As the document says: "all of (these points) can be implemented immediately and sustained after this particular crisis has subsided."

If efforts at change are to move beyond mere bandages to recurrent injuries, institutional practices must change. Those in the Netherlands who support the five points in the manifesto recognize this and are working to gain political backing. The same is true for supporters of the Green New Deal (GND) in the U.S. Eventually, backers of the GND will need to confront the necessary, but difficult, transitions that are explicitly recognized in the Netherlands document.

Another international group, ***Feminism for the 99%*** (Arruzza et al. 2019), has issued a manifesto analyzing current crises. The basic crisis we face, they state, is one of society as a whole, one rooted in capitalism. The manifesto offers a unifying goal: we need to place people-making over product-making for profit. We must turn our values and priorities right side up. People are the priority; people-making involves not only the labor involved in social reproduction (the creation of people, their care and education) but also the labor that goes into imparting skills, values, and knowledge throughout life. Capitalism is based on, and maintains its power because of, divisions among us. Overcoming these divisions is key to the solidarity society. Divisions may be based on culture, race, ethnicity, ability, sexuality, or gender. *Feminism for the 99%* issues a series of demands that address the stratification of society, the violence against women perpetrated under capitalism, and the undervaluation of unpaid and caring labor. It calls for the liberation of sexuality and a reversal of capital's destruction of the earth. By uniting in a movement to "turn values right side up," change is possible.

The Care Manifesto, issued by the Care Collective of London, has a similar message.[39] It posits that care be the organizing principle of every dimension of life: care for one another and for the earth. The economy, the state, and the home should strive to create collective joy.

Another call for change, authored by several eminent scholars and signed by thousands more, calls for democratizing work, decommodifying labor, and remediating the environment [http://democratizingwork.org/].[40] This document shares the concerns of the Netherlands group, but places focus on democratizing work and increasing people's voice. "Caring for the sick; delivering food, medication, and other essentials; clearing away our waste; stocking the shelves and running the registers in our grocery stores – the people who have kept life going through the COVID-19 pandemic are living proof that work cannot be reduced to a mere commodity," the group notes. Greater democratic control of business with new priority on the public interest would bring lasting changes in political campaigns, in the laws and rules that now stifle collective bargaining, financial practices that disadvantage small businesses, farms and minority groups, and more. The call to democratize, decommodify, and remediate is influencing academic research and teaching, and inspiring activists in the field.

Environmental groups including the Sierra Club, Greenpeace and 350 have broadened their approach to protecting the environment to include steps toward advancing racial justice and protecting the rights of workers, women, Indigenous peoples, and other minorities, recognizing that everyone needs to be involved in saving the planet, and that this is only possible when everyone's urgent concerns are being addressed.[41]

The Movement for Black Lives, which includes Black Lives Matter as a coalition member, is a group which has grown in influence after police killings in the U.S. in 2020. It offers an analysis of institutionalized racism and a series of demands that address racism throughout society. It calls for an end to injustices in police and criminal justice systems. It supports the Green New Deal. It calls for investment in people, and divestment from prisons, fossil fuels, police, surveillance, and exploitative corporations. It calls for community control of what's happening in the community, and offers recommendations for greater political power for people and for economic justice, including in regard to the response to COVID-19. Local branches can add their own locally directed demands.

There are thousands of groups striving to create a better world for themselves and others. Besides the five movements we have focused on here—labor, women's rights, racial justice, environmental, and LGBTQ+ —there are immigrant rights groups, disability rights, and many more.

The ***World Social Forum***, founded in 2001, is an open meeting space for reflective thinking. It is committed to "a globalization based on the values of ***solidarity*** that, as a new stage in world history, will respect universal human rights and all citizens…of all nations, as well as the environment and will be underpinned by democratic international systems and institutions at the service of social justice, equality, and national sovereignty."[42] The civil organizations attending these forums, which are held in different countries around the globe, debate and exchange ideas on the creation of new local and world systems. Non-Governmental Organizations (NGOs), including many funded by foundations, have significant presence and voice. Organizations that cannot afford to send a representative to the international meetings may exchange ideas at regional events. For a while the meetings were annual but frequency of meetings has declined.

There are also affiliated organizations in several countries, including the U.S., where a Solidarity Network has been established.[43] These networks provide valuable connections and opportunities for the exchange of ideas and collaborative movement-building. Like the World Social Forum, they are not set up to take political action.

In addition to organizations working over the long run for change, there have been huge *demonstrations* around the world pressing demands for change. Some noteworthy protests of 2020 included: Black Lives Matter; protests against specific killings of Blacks, LGBTQ+ persons; climate; women's day; anti-special police violence in Nigeria; and pro-democracy in Hong Kong. There is political energy in abundance, as more recent events also show.

The Critical Challenge: Creating Unity

Today's movement groups, no matter how committed and progressive, focus on a few issues they have chosen and seldom mention coalition work as a priority. When we examine the manifestos and political agendas above, we find that no one manifesto or document addresses key concerns of each of the five major movements (labor, feminist, racial justice, LGBTQ+ , environmental) in detail. Movement organizations may share the same values, but their agendas, although radical in some areas, are distinctly different.[44] The theoretical underpinnings, be they sophisticated and well-thought-out or more pragmatic and traditional, differ as well.

Currently, progressive organizations occasionally unite in support of specific pieces of legislation such as the Build Back Better bill under Biden. But even important bills like the PRO Act have not received much support across organizations. [The Protecting the Right to Organize Act (PRO) would remove roadblocks to union organizing and greatly increase employees' collective voice at work.] This and other important bills are under-reported in the news, and are not receiving the strong, unified support from progressive groups they need.[45] However, most activist groups have education and outreach as part of their mission. It is strategically wise that their outreach extend to other progressive groups. The nature of this networking would be decided within each group, but such effort is basic to creating the unity needed for progressive change.

Unity among disparate movement organizations would be greatly strengthened by consciousness-raising within each organization. Leaders as well as members of these organizations should know the basics about racism, patriarchy, other forms of discrimination, imperialism, the role and place of the military, climate change, and the environment. Understanding needs to grow, as Yates (2018), Matthaei (2017), and others discuss. Without this, internal divisions are likely in coalitions and success will be limited accordingly.[46]

This educational effort requires more than simply having members listen to a few talks. Racism, sexism, and other forms of oppression are systemic, and an understanding of how social systems work to the disadvantage of certain groups requires both in-depth study and on-going involvement.

However, the internal transformation of every movement and progressive group is not a pre-requisite to building a left populism. In fact, left populism is possible without such in-depth transformation. Certainly, movement organizations should have a commitment to learn about and support the radical projects of other movements. An appreciation of the transformative nature of each group's radical projects is also important.

But if all groups accept that debate within the coalition they are building will be respectful, open, inclusive, and democratic, successful coalition work will be possible. Importantly, each group will have an incentive to gain the support of other coalition members when working for the transformed society they envision. This support will be contingent on acknowledging and addressing any shortcomings pointed out by fellow coalition members. Working together well requires meaningful communication and learning. The standpoints of minorities—especially those suffering intersectional oppressions—must be sought out, if there is to be clarity of purpose. There is a common interest in a democracy that recognizes and addresses the critical issues raised by today's movements.

This common interest in deeper democracy, and in saving the planet, can be the basis on which to build a left populist coalition.

Successful coalition-building, then, depends critically on coalitions adopting decision-making processes that give adequate representation to each coalition member. Each member should find that membership helps it to advance its goals. Creating strong democratic processes must be done in a deliberate fashion. Organizations come into coalition, not only with different viewpoints, but also with different resources and membership numbers. The relative strength of the different groups in the coalition will strongly influence the starting agenda from which further gains can be made. A coalition that believes in equity will acknowledge these inequalities and take steps to address them. Rules should ensure that those with greater wealth do not dominate or disrupt debate. The rights and voice of minorities must be protected and given priority when there have been egregious violations of their rights by society—such as the separation of children from their parents at the U.S. border as has occurred. The funding of research projects addressing the priorities of under-funded groups can also help to advance equity and justice within the coalition. True interests theory lays out some of the pre-requisites for a group to recognize, articulate, and self-organize for their interests.[47]

Building a better future demands working together. One simple step toward networking would be for an organization to inform their membership about noteworthy pieces of legislation being considered, bills that have the strong backing of progressive movement groups. The organization need not even take a position on that legislation—just a brief description taken from the bill would educate members.

Right now, groups do not appear ready to agree to even a general agenda. A coalition that included major organizations from all of the five social movements and coalesced around a broad agenda would mark a significant step forward. It would undoubtedly achieve notable success. Such an agenda might include, for example: ensuring that all women have access to safe and legal abortions, reducing carbon emissions to net zero by 2050, redirecting funding from police to increased funding of social services, and ensuring that workers' right to unionize is protected. Affordable child care and health care and other priority demands are also potentially part of progressive political agendas, but as of now any coalition work around such issues remains scattered. Movement groups remain largely specialized, and divided.

On the populist path to a deeper democracy, demands for policies that increase the voice and power of Middle America and disadvantaged groups are of particular importance. Movement organizations can lead by example. Patriarchy, racism, and other ills are being tackled within organizations. The non-democratic and oppressive conditions at workplaces have long been addressed by unions. Solidarity initiatives that support organic farming, local currencies, affordable housing, cooperatives of all sorts, farmer's markets, immigrant and prisoners' rights and much, much more are fostering a belief in the possibility of democratic economies. Efforts to establish consumer-owned utilities across the U.S. are strengthening the ability of citizens to challenge corporate control of the electrical power grid and to gain control of the future of energy in their regions.[48]

These efforts are dispersed, diverse, and in great need of intersectional solidarity.

Fiscal Feasibility of Programs Above

A restructured economy, such as those described in the manifestos above, requires restructured financing of government. Luckily, such economies would benefit from positive dynamics: thriving economies are good for government coffers. The higher earnings for workers mean a reduced need for government assistance programs, like food stamps. Higher incomes also mean that individuals can afford to pay income-adjusted fees for child care, health care, and other services. Higher taxes on the wealthy would be more politically feasible. The added funds would enable the investments needed for the transition to a green economy, from green infrastructure to research and education. These investments have positive and substantial rates of return. Any increased taxation of the middle class would be more than offset by reductions in personal spending on college tuition, health care, child care, elder care, transportation, and more.

A middle class on more solid ground would be better positioned to pursue higher education, to start a business, or to spend more time pursuing goals in volunteer work, music, art, or other interests. Greater employer-employee-community cooperation would reduce the time and energy needed to enforce compliance with regulations. Time spent by business in battling competitors for market share would drop as the pursuit of profit lost appeal, and more business energies could be used cooperatively to solve shared challenges. The restructuring of the economy would mean a shift in defense spending toward international assistance and loans and a reduction in unnecessary military spending: the U.S. currently spends more on the military than the next nine top-spending countries combined.[49]

More people would have healthy diets, exercise and get enough sleep as access to these basics improved. Violence against women would be reduced as attitudes change and women are assured of the housing and income they needed to leave abusive relationships. Access to treatment for drug addiction and mental health issues would prevent the escalation and spread of these social problems. And, of course, the incidence of these and other problems arising from oppressive attitudes would decrease under a more caring society. With more equal opportunity, society's talent pool would enlarge, and creativity and innovation would increase. These developments would contribute to higher output per hour of work.

Tax loopholes would be closed, and new tax brackets added for the highest income groups. New sources of revenue would be explored. Governments could require royalty payments when government-funded research resulted in higher earnings for a firm. Loans would be made more easily available to smaller businesses if banks included local economic development as one of their priorities. Publicly or community-owned banks would grow in importance and local development would be an honored goal. In addition, the profits of these institutions could be a funding source for low- to moderate-income housing which sometimes offers only a modest rate of return.

There would be significant amounts of time saved. Technological change would be welcomed as bringing higher pay and fewer hours of work. More worker rights would mean fewer middle-level managers—the now bloated ranks of overseers would shrink, in line with the observations of some business experts.[50] People would spend less time in traffic if there were shorter workweeks and no need to work at 2 or 3 jobs. Less time would be spent searching for care providers for dependents and filling out forms. Less planned obsolescence would reduce shopping time, and more locally sourced agricultural products would reduce transportation needs. Greater employee earnings would shift market production toward goods and services demanded by the bottom 90%, and away from the risky investments in financial assets and real estate ventures favored by the top 1–5%. More effective government oversight of banking would also lead to greater monetary stability and an end to widespread financial crises due to risky investments.

Data show that greater investment in people and the planet is immediately feasible with a redistribution of income and wealth. We now produce more than seven times as much per person as we did in 1900 and less than 2% of the workforce is in agriculture and less than 10% in manufacturing. Output per worker hour is twice as high as it was in 1980, but hourly wages remain almost the same as in 1979. A redistribution of output and income is clearly feasible and desirable. If income generated in the U.S. were equally divided, a household of four would receive over $260,000.[51]

Alternative Motivations at the Heart of a Solidarity Society

Motivation for economic activity varies. The profit motive and personal gain are the incentives driving economic activity, at least according to many economists. This message is repeated in classrooms across the country and spread by the media. Fortunately, it is a message that is wrong. Although the drive for profit or gain has considerable explanatory power in one sense, a desire for higher income and wealth in fact is not a primary motive for human behavior. Rather, it is a derivative motive, a means to an end. This is an important distinction: it means that people are driven by goals other than by money per se, and that there are other motives for work effort and other ways of incentivizing economic production. People work toward goals within the social framework in which they are placed.

Think of our current economic system as a competitive game called capitalism. Each player begins with different resources and faces choices that lead to different outcomes. Some people play to get to the top of their league; others play simply to get by. But we're all in it, like it or not. It is a game set up long ago, but with rules that are always changing. Those at the top of the game are the ones who can change the rules, although from time to time coalitions of other players can force the adoption of new rules that improve their odds of doing well. How would a solidarity society motivate people to play "democratic economy"—what would this new game be like?

It is worth noting that most economic activity today is not driven by cut-throat efforts to get to the top. The majority of people are employed in the public sector, in small businesses, and in nonprofits.

Behavioral psychologists classify the primary motives behind human behavior according to human needs and desires. Although variously articulated, as we saw earlier these include:

1. needs related to survival and reproduction,
2. the desire for safety, health, and security,
3. social needs: the need for acceptance and recognition, and for fulfilling relationships. These involve commitment and shared values,
4. the individual need for self-actualization (which spurs the creative drive to discovery, self-expression, independent values, enjoyment of pleasure and play).

These four are inseparable and overlap—for example, as individuals honor their personal and social commitments, whether to family, country, or community, they act in ways that enable them to strengthen their personal ties and sense of self-worth, but they also gain social recognition. When the commitment is to a marriage, the sexual drive and the need for safety and security are also addressed.

Contrast this list of human motives with the motive singularly emphasized by economists: the pursuit of economic gain. Public policy recommendations are built on the assumption that economic rewards and penalties are the primary, preferred means of motivating human effort. Reconsideration of this assumption opens the door to new policy options. Motives besides personal gain, in fact, are often the drivers of social behaviors. In wartime, a sense of duty to one's country motivates people to enlist in the military. A person's occupational choice is seldom based on a desire to maximize income, but rather on a desire for fulfillment and for income adequate to ensure a desired standard of living. Priorities and values impact choice: people usually try to do the right thing as they see it. They put caring for loved ones and protecting their communities above seizing every opportunity for gain. If a given increase in income will bring only marginal change to a person's life, the added income might have less appeal than the satisfaction derived from meaningful volunteer work or from spending more time with family and friends.

The ruthless pursuit of profit that dominates the economic behavior of the few at the top of the economic ladder is the inevitable result of the compulsion to be winners in the capitalist "game." Those at the top are in it to win it. The high-stakes game that "we" have crafted over the years has had unfortunate consequences. It has made the U.S. an increasingly dysfunctional society, one in which the expedient choice too often wins over the socially responsible one. Businesses willingly accept high levels of risk and negative "side-effects" on human rights and the environment for the sake of higher returns. If we want the more socially responsible decision to be chosen over the more profitable one, the goal of monetary gain must drop in importance relative to other goals in the business world. This is not as unlikely as may appear.

Business has adapted to many rule changes over the course of time and will continue to do so. The laws and regulations likely to come up for consideration under a radical democracy would aim to raise worker earnings, improve services to the population, and provide greater protection of the environment. These changes might reduce net income accruing to those at the top of the income ladder, but a more economically secure workforce and sounder ecological practices would bring great benefits. In some industries, competition from countries that do not enact such changes could tilt the competitive playing field but, in larger countries, any resulting problems could be addressed through modifying trade and investment agreements.

Motivations will play a critical role in the political effort to change the rules of the economic game. Will there be a political party motivated by a vision of a better world? There always has been. To be effective, a transformative party will draw on shared values and demand the commitment that comes from strong beliefs. The goal of shared prosperity and a healthy environment can replace individualistic goals of personal gain.

Political analysts have long been clear that a person's commitment to a political party or candidate depends on the political identity that person has worked out for him or herself. After Trump's election to the presidency and the rise of right-wing political factions around the world, the role of affect (emotional connection) in political choice has received more attention from the left. A person's political identity emerges from interactions with different political groups. The group that becomes the individual's political home will be the one that shapes that individual's political positions. Leadership and structure matter: positive energies can be harnessed for good in well-structured organizations, just as negative energies can be harnessed and misused. Self-organization, clarity, and mutual respect move a political party's progressive agenda forward. Political debate with real give-and-take and regard for relevant data is the norm in many organizations; it is behavior that can be learned.

In the economy, the drive for profit and income can wane in importance as economic security becomes widespread. Once a decent standard of living is assured for all regardless of occupation, once everyone can afford the education they need to pursue the career or vocation of their choice, new channels for creativity and connection will be opened. Motivations other than material gain will rise in importance and individuals will gain respect and stature in their communities from their creativity, kindness, courage, and achievements in their chosen fields. This recognition from peers and others can rival or eclipse the attraction of monetary wealth. It will take time and self-reflection within society's many institutions before human worth as opposed to monetary worth becomes the norm by which people are valued.

Conclusions

The wealthy are powerful—their power shored up by their domination of social institutions and enhanced by ideological messaging spread through the media. Their wealth and power derive from income received from their business and other investments. They control wealth that is created collectively: by scientists, researchers, engineers, technicians, assembly line workers, sales personnel, and others who do the essential work. These wealth-creators, however, typically have little or no say in the use of the wealth they generate. This happens only with society's consent!

Any challenge to the domination of our democracy by plutocrats will at the same time, then, be a challenge to the rights now attached to the ownership of wealth. The future of our planet depends on addressing climate change. Plutocrats do not want to take the steps needed to prevent catastrophe, steps that would irrevocably take away their current privileges. A rapid end to the use of fossil fuels would throw millions of people into dire poverty—unless resources currently controlled by the wealthy are dedicated to providing people the support needed during the transition to clean energy. Those of us in high-income countries count on fossil fuels for transportation and home heating; products we buy are produced with fossil fuel inputs. Challenges are similar in lower-income countries. Massive investments, not only in clean energy, but in people are critical to enabling all to survive and thrive during the transition to an eco-friendly world. Popular support for this new direction depends on the transition being done with deeply democratic input and an eye to improving people's lives as well as the health of the planet.

Businesses—both large and small—will play a major role in this change. They are part of society and society has the right to expect that business operations are carried out in line with shared commitments to human rights, sustainability, and to practices that enable planet and people to flourish. Many business leaders recognize their social responsibilities, have re-envisioned the role of their businesses, and are remaking them accordingly.

A better future depends on transforming how we do business. A responsible and responsive business sector that is accountable to the broader community is a goal that is, not only feasible, but urgent. A shift of power within our giant companies—from a small, wealthy elite to a balanced representation of labor, communities, consumers, and other stakeholders—is critical to decision-making that would ensure a prosperous, eco-friendly future.

There has always been push-back against the abuse and misuse of power by those controlling production. Slavery has long been illegal. Laws against child labor, against unnecessarily dangerous work conditions and against unpaid hours of work have eventually made it through Congress and been upheld by the courts—if only after decades of struggles by workers and their allies.

A firm commitment to the public good demands leadership that truly represents the people. When and if a radical democracy is established, new norms, practices, and expectations will be born. Efforts will be focused on saving the planet, ensuring individual well-being, nurturing creativity, and strengthening community ties. New laws and practices will aim at ending the lower status now plaguing many groups.

The political struggles that can move us toward this solidarity society have begun. We are in a period of turmoil—significant political shifts are underway. We face both potential and peril. During the past 100 years, workers and their families were able to overcome enormous obstacles in their struggle to improve their economic, social, and political status. Many paid a high price, but their successes continue to benefit those who came after. Current struggles are a continuation of those earlier ones. Social movements—labor, racial justice, environmental, LGBTQ+, women's liberation, disability rights, and more—can form the base of a political coalition capable of instigating social transformation.

The creation of a deeply democratic, caring solidarity society, in the end, rests on the hope, love and courage of those dedicated to its realization. It is past time that today's emphasis on material wealth as the measure of value, even human value, gives way to more humane understandings. Millions of people around the globe recognize this and millions will be taking part in political battles, demonstrations, and social movements in the coming years—efforts that are inspired by the conviction that the good can and must prevail. This is nothing new—across time and place there have always been people who have thrown their lot in with others fighting for a cause that resonates deeply.

Those who are dedicated to sustainability and justice have difficult battles to fight. The status quo is everywhere protected by the prevailing hegemony and by the ideologies that uphold it. Those in power plan to keep an upper hand. Only grassroots mobilizations that have a concentrated focus on ending the domination of the wealthy few will be able to change the values and practices that steer our capitalist societies.

The institutions and mechanisms that protect the accumulation of wealth need to be replaced by democratic practices that bring collective control of collectively created wealth. The unified coalitions required for this work will be built through democratic processes that honor shared values and goals. One key challenge in these efforts is finding platforms from which to reach out and connect. Workplaces are fragmented; interactions among different communities are limited. People may have a deep yearning to be part of an expansive, caring state- or nation-wide community, but they are divided by racism, sexism, and other oppressive ideologies. In addition, organizations dedicated to social good are often too narrowly focused in their work. Many have yet to recognize the urgency of working together in coalitions that embrace the radical projects of other movement groups.

It is past time for progressives to place priority on networking that is supportive of all of today's struggles for justice and sustainability. Such cooperation would lead to powerful coalitions that can move us toward a new society. These efforts must rest on shared values and a shared commitment to respectful, open, democratic processes, not on agendas determined a priori by a few. Through such processes, the way forward will emerge.

People have always managed to find ways of building the unity of purpose required to achieve a shared goal, even 100 and 200 years ago. Today's barriers to unity may be different, but this only means that ways of overcoming the new obstacles must be created. Mass media, for one, presents a formidable challenge, but it is not impenetrable. The growth of left populism in Britain, the U.S., and elsewhere is proof that creative thinkers have found ways to connect and unite. The powerful few will only yield power to a strong and effective opposition. Proponents of a caring and sustainable solidarity society are many and, once united, can be that effective opposition. Clarity, cohesion, and courage will be enough to ensure success.

Appendix: Democratic Oversight of Business—Some Options with Teeth

Federal charters: This was discussed above as an option for a more democratic economy. Firms above a certain size would be required to be federally chartered and subject to penalties for violating federal laws and regulations. Loss of limited liability—a possible penalty—would place the assets of individual owners at risk and so increase responsible behavior. The downside is that the firm's access to funding for investments would drop sharply. Less severe penalties could include court orders that would require compliance with sanctions under penalty of severe fines, for example. Sanctions could include requiring that regulators have access to accounts or be present at board meetings.

Independent financial and social auditing (not paid directly by the firm being audited but by fees paid into a common fund). Social auditing would report on wages and benefits paid, progress in representation of women and minorities in workforce and management, impact on the environment and community, and similar.

Require that a majority of corporate board members in large firms represent the public interest. Businesses should become eco-friendly and socially responsible. The choice of persons to represent the public on business boards should be through a democratic process that involves grassroot organizations. These organizations, and local municipalities, could nominate potential corporate board members who, after being approved, would be trained for a year at university business schools. A representative democratic body such as a state legislative committee could be the body choosing the candidates for training from among the nominees. The training could be financed by a small tax on shares, derivatives, and options traded on exchanges.

The requirement that a certain percentage of board members be professionally trained and independent of the corporation involved is an idea that has long been proposed by business consultants as a means of improving corporate boards. But rather than corporations choosing the professionals to be on their boards, their choice could be limited to democratically approved candidates.

Turn ownership of the firm over to a foundation. Colin Mayer (2013), former Dean of the Oxford University Business School, envisages this as the path to corporate reform, with shareholders voluntarily setting up nonprofit trusts to oversee corporate activity. Some of the largest companies in the world are majority-owned by foundations, typically set up by former family owners. These include Bosch (the German electronics firm well-known for its automotive components), the Tata Group (an Indian conglomerate that bought out British Steel and Jaguar), and Novo Nordisk (a Danish pharmaceuticals firm). The foundations or trusts provide oversight of the overall conduct of business, ensuring that the firm adheres to founding values and honors its commitments to stakeholders. Democratic control of such trusts is a possibility that would serve the public interest.

However, only a business culture that places the public good ahead of private gain will prevent the need to devote extensive resources to oversight and regulation. Employees, communities, and customers must have significant voting representation on the key committees and boards of companies. This would bring effective democracy to business. And dealing with systemic problems like discrimination and climate change will require further radical reforms, including, for example, a participatory planning process. See Hill (2014).

Notes

1. https://www.cnbc.com/2019/09/19/jeff-bezos-speaks-about-amazon-sustainability-in-washington-dc.html.
2. *Doughnut Economics* (Raworth 2017). *A Finer Future* (Lovins et al. 2018).
3. Lovins et al. (2018: 321). Hill (2007) addresses the biases often found in the reform efforts of international organizations.
4. According to the definition used by the Levy Institute, "financialization is a process whereby financial markets, financial institutions, and financial elites gain greater influence over economic policy and economic outcomes. Financialization transforms the functioning of economic systems at both the macro and micro levels." The rise of financial officers in firms and the growth of collateralized debt obligations are part of this process.
5. https://www.ipcc.ch/report/ar6/wg2/.
6. Council of Economic Advisers (2021: 494).
7. Nussbaum (2004: 49).
8. U.N. Universal Declaration of Human Rights is at https://www.un.org/en/about-us/universal-declaration-of-human-rights.

9. See e.g. Wright (2019), Matthaei (2017), Yates (2018: 141–143). Sweden is among the countries and cities that have set ambitious goals: https://platform2020redesign.org/countries/sweden/.
10. OECD http://www.oecd.org/statistics/how-s-life-23089679.htm.
11. Hill (2015: 169).
12. http://fairvote.org/.
13. Hill (2015: 165). Also Hill (2017).
14. Hill (2014). Mondragon Cooperative, headquartered in the Basque region of Spain, is a highly successful federation of worker cooperatives, with over 80,000 persons employed in over 200 companies ranging from research to finance and several manufacturing industries.
15. Alperovitz (2005: 73). See also Alperovitz (2013) and (2017).
16. Yates (2018: 113–114) and Bergstrom (1992: 162–170). Rudolf Meidner's proposal for establishing Wage-Earners'Investment Funds was presented for adoption by the Swedish Trade Union Confederation (LO)'s national congress in 1976. The original idea was that gradually workers would gain majority ownership, but that goal changed to holding some shares in the business. A joint LO and Swedish Social Democratic Party (SAP) committee wrote a recommendation presented in 1981 to both the LO and the SAP. In 1983, a much more modest version of Wage Earners' Investment Funds was established. The funds were privatized in 1991. Wage Funds remain a controversial topic in Sweden.
17. Another approach to upholding a firm's commitment to its stakeholders would be to turn over ownership of the firm to a foundation set up to oversee the firm. Some of the largest companies in the world are majority-owned by foundations, typically set up by former family owners. These include Bosch (a German electronics firm), the Tata Group (an Indian conglomerate that bought British Steel and Jaguar), and Novo Nordisk (a Danish pharmaceuticals firm). But the values and commitments upheld by the trust are determined by the persons setting up and controlling it—only a trust under democratic control would have truly democratic values.
18. A financial system that is socially responsible would do its part to ensure that it does not finance activity that is detrimental to society. This could involve, for example, banks checking the environmental impact of a project before granting funding. The social effects of a project on the employment and the work loads of women and minorities could also be assessed as part of the loan approval process. Already, some financial institutions such as publicly-owned economic development banks are explicitly charged with doing more than pursuing the interests of shareholders and customers. At a global level, the World Trade Organization enforces international agreements and the International Monetary Fund imposes sweeping conditions on some of its loans. The power of these international organizations,

however, is exercised in the context of expediting smooth functioning of today's economic system.

19. Alperovitz (2017) argues that to delegate authority to regional bodies a change in the U.S. Constitution would be required. See his concluding remarks. Examples of eco-friendly alternative businesses, promising public–private initiatives, local goods and services exchanges and more can be found at the website of the Next System: https:thenextsystemproject.org/principles-introduction Juliet Schor (2011) also gives a wealth of examples of grassroots efforts creating a better world.
20. https://www.euronews.com/green/2022/02/28/today-s-ipcc-report-proves-green-capitalism-doesn-t-work-says-climate-expert. The report and summary are at: https://www.ipcc.ch/report/ar6/wg2/.
21. https://data.worldbank.org/indicator/SP.POP.GROW.
22. https://www.ipcc.ch. An important resource from noted scientist and climate activist Paul Hawken is https://regeneration.org/. The tagline reads: "The world's largest, most complete listing and network of solutions to the climate crisis. And how to do them!" For insight into how to get legislation passed at the state and local level, see Inches (2021).
23. https://www.space.com/17683-earth-atmosphere.html and https://www.nationalgeographic.org/article/greenhouse-effect-our-planet/4th-grade/.
24. Annual emissions of the greenhouse gas rose by 9% 2000–2017, which has the same effect on global warming as adding 350 million cars to our roads. Cattle ranching is the main reason for increased methane levels. Methane is 28 times more effective than carbon dioxide at trapping heat over 100 years."Emissions from cattle and other ruminants are almost as large as those from the fossil fuel industry for methane," Jackson said. "People joke about burping cows without realizing how big the source really is" https://www.cnn.com/2020/07/15/world/methane-emissions-record-scli-intl-scn/index.html.
25. https://eos.org/articles/reindeer-could-trample-permafrost-thaw.
26. https://mronline.org/2007/02/28/the-swedish-welfare-state-a-model-for-canadian-labor/, https://www.iea.org/reports/cementhttps://www.weforum.org/agenda/2019/09/cement-production-country-world-third-largest-emitter/ reports on several new technologies https://www.rechargenews.com/transition/vattenfall-plans-world-s-first-zero-carbon-cement-plant/2-1-531250tomorrow-documentary.com. This excellent documentary covers a lot of ground. Note the focus on ensuring the well-being of all. Some carbon tax advocates fail to attempt to redress the hardship that can accompany a shift from fossil fuels. But the funds from such a tax could be used to address job loss, a rise in the cost of home heating and in transportation costs, and similar.

27. https://www.defense.gov/News/News-Stories/Article/Article/2787056/dod-announces-plan-to-tackle-climate-crisis/.
28. See, for example, reports of the Mississippi Immigrant Rights Alliance (yourmira.org) and the book by David Bacon (2008).
29. Matthaei (2017: 147–57).
30. https://thenextsystem.org/principles-introduction. See conclusion.
31. Yates (2018: 144).
32. Wright (2019: 37–64).
33. https://www.britannica.com/topic/critical-theory.
34. Mouffe disagrees with Hardt and Negri on this point. See endnote below.
35. Mouffe (2018). While Mouffe advocates vertical engagement, some on the left believe that the people/multitudes will spontaneously organize in ways that make involvement with mainstream politics optional (horizontal organizing). See her critique of M. Hardt and A. Negri, pp. 53–55.
36. MacKinnon (1993).
37. House Resolution 1102, U.S. Congress September 11, 2020. https://www.congress.gov/116/bills/hres1102/BILLS-116hres1102ih.pdf.
38. https://ontgroei.degrowth.net/manifesto-for-post-neoliberal-development-five-policy-strategies-for-the-netherlands-after-the-covid-19-crisis/.
39. Care Collective (2020). From the publisher's description: "We are all dependent on each other, and only by nurturing these interdependencies can we cultivate a world in which each and every one of us can not only live but thrive." https://www.versobooks.com/books/3706-care-manifesto.
40. https://democratizingwork.org/. Work: Democratize, Decommodify and Remediate.
41. https://www.bluegreenalliance.org/the-latest/bluegreen-alliance-announces-new-executive-director/.
42. https://transformadora.org/en/about/principles.
43. https://ussen.org. See also Benner and Pastor (2021).
44. In comparing the manifestos, platforms and other statements of the groups mentioned above and a few more, I found that while there appeared to be shared values, I was unable to find any specific goal common to all groups. For example, although the goal of net zero greenhouse gas emissions by 2050 is likely a goal shared by all environmental groups, it was not easy to find statements confirming that assumption. Non-environmental movement groups would occasionally mention support for sustainability or clean energy or an end to reliance on fossil fuels in passing, but environmental goals were seldom listed as one of their issues. Support for reproductive rights was also often neglected. It is likely that the leadership of most progressive groups does in fact recognize the urgency of ending climate change or protecting reproductive rights, but these groups do not always provide their members with a means of accessing information on

these issues. If a group is working in coalition with one or more environmental groups, it is likely around specific legislation, and is not an on-going partnership. More generally, goals of one movement group did not often include specific goals associated with other movements.

45. Progressive organizations that are not focused primarily on labor issues seldom mention the PRO Act on their websites (based on my internet searches of several large activist organizations). I also checked whether activist websites expressed support for women's reproductive rights. I found few non-women's organizations that did.
46. Yates (2018: 146–150) presents reasons why education in such areas is critical to coalition-building.
47. Hamilton (1999). Inclusivity in decision-making does not mean simply including voices of those who have historically been excluded. Just as choosing someone at random off the street and asking her opinion about a proposed law affecting women is not the way to find out how women would be affected by that law and what they would prefer. Research on the proposed bill and its impact on women must be available if those representing women are to have an informed opinion. The source of this information is also an issue; and several sources are preferable to one.
48. See ourpowermaine.org for discussion of such an initiative in Maine.
49. U.S. spending on the military was $778 billion in 2020; China spent about $252 billion and Russia, $62 billion. The military-industrial complex here exerts enormous influence in Congress. https://worldpopulationreview.com/country-rankings/military-spending-by-country.
50. https://hbr.org/2016/09/excess-management-is-costing-the-us-3-trillion-per-year. Economist David Gordon made a similar point over 25 years ago, stressing the workers' perspective on intrusive oversight.
51. *Economic Report of the President 2021*, Appendix B. 2019 data.

References

Alperovitz, Gar. 2005. *America Beyond Capitalism*. Hoboken, N.J.: Wiley.

———. 2013. *What Then Must We Do?* White River Junction, VT: Chelsea Green.

———. 2017. *Principles of a Pluralist Commonwealth*. Democracy Collective. https://thenextsystem.org/principles

Arruzza, Cinzia, Tithi Bhattacharya, and Nancy Fraser. 2019. *Feminism for the 99%*. London: Verso. https://outraspalavras.net/wp-content/uploads/2019/03/Feminism-for-the-99.pdf.

Bacon, D. 2008. *Illegal People: How Globalization Creates Migration and Criminalizes Immigrants*. Boston: Beacon Press. Chapter 6 on Mississippi and Katrina.

Benner, Chris, and Manuel Pastor. 2021. *Solidarity Economics*. New York: Wiley. Open source copy at: https://solidarityeconomics.org/se-book/.

Bergstrom, Villy. 1992. Party Program and Economic Policy: The Social Democrats in Government. In *Creating Social Democracy*, ed. K. Misgeld, K. Molin, and K. Amark. University Park, PA: Pennsylvania State University Press.

Care Collective. 2020. *The Care Manifesto: The Politics of Interdependence*. London: Verso.

Council of Economic Advisers. 2021. *Economic Report of the President 2021*. Washington, DC: U.S. Government Printing Office. Appendix B.

Democratizing Work. A manifesto at https://democratizingwork.org/.

Hahnel, Robin. 2012. *Of the People, By the People: The Case for a Participatory Economy*. Soapbox Press.

Hamilton, Lawrence. 1999. A Theory of True Interests. *Government and Opposition* 34 (4): 516–546.

Hawken, Paul. 2021. *Regeneration*. London: Penguin. See https://regeneration.org/.

Hill, Marianne. 2007. Confronting Power through Policy. *Journal of Human Development and Capabilities* 5 (2), July.

———. 2014. Taming the Corporate Beast, Part 2. *Dollars and Sense*, November. Boston: Economic Affairs Bureau.

———. 2017. Taming the Corporate Beast. In *Imperiled Economies 2018*, ed. URPE Reader Collective, 139–146. Boston: Economic Affairs Bureau.

———. 2021. Creating an Economy that Prioritizes People and Planet. Book Review. *Dollars and Sense*, March. Boston: Economic Affairs Bureau. http://www.dollarsandsense.org/archives.html.

Hill, Steven. 2015. European Democracy's Slow and Patient Rise. In *Lessons from Europe?*, ed. R.D. Kelemen, 163–186. Los Angeles: CQ Press.

Inches, Susan B. 2021. *Advocating for the Environment*. Berkeley: North Atlantic Books.

Lakey, George. 2017. *Viking Economics: How the Scandinavians Got It Right and How We Can, Too*. Brooklyn: Melville House.

Lovins, L., S. Hunter, A. Wallis, and J. Fullerton. Wijkman. 2018. *A Finer Future: Creating an Economy in Service to Life*. Grabiola Island, Canada: New Society Publishers.

MacKinnon, Catharine A. 1993. Feminism, Marxism, Method, and the State: Toward Feminist Jurisprudence. In *Feminist Legal Theory Foundations*, ed. D. Kelly Weisberg, 427–36. Philadelphia: Temple University Press.

Maslow, Abraham. 1972. *The Farther Reaches of Human Nature*. London: Penguin.

Matthaei, Julie. 2017. From Inequality to Solidarity. In *Imperiled Economies 2018*, ed. URPE Reader Collective, 147–158. Boston: Economic Affairs Bureau.

Mayer, Colin. 2013. *Firm Commitment*. Oxford: Oxford University Press.

Mouffe, Chantal. 2018. *For a Left Populism*. London: Verso.

Nussbaum, Martha C. 2004. Capabilities as Fundamental Entitlements. In *Capabilities, Freedom and Equality*, ed. B. Agarwal et al. New Delphi: Oxford University Press.

Ontgroei (Degrowth). April 16, 2020. *Planning for Post-Corona: Five Proposals to Craft a Radically More Sustainable and Equal World*. The Netherlands. https://ontgroei.degrowth.net/manifesto-for-post-neoliberal-development-five-policy-strategies-for-the-netherlands-after-the-covid-19-crisis/.

Organization for Economic Cooperation and Development (OECD). Various years. *How's Life*. https://www.oecd.org/sdd/How-is-Life-2020-Highlights.pdf

Pickett, Kate, and Richard Wilkerson. 2009. *The Spirit Level*. New York City: Bloomsbury Press.

Piketty, Thomas. 2022.*A Brief History of Equality*. Cambridge, MA: Belknap Press.

Raworth, Kate. 2017. *Doughnut Economics: 7 Ways to Think Like a 21st Century Economist*. White River Junction, VT: Chelsea Green Publishing.

Schor, Juliet B. 2011. *True Wealth*. New York: Penguin Group.

Shutz, Eric A. 2011. *Inequality and Power*. New York: Routledge.

Stiglitz, Joseph. 2020. Measuring What Matters. *Scientific American*, August, 24–32.

Tronto, Joan. 2013. *Caring Democracy: Markets, Equality, and Justice*. New York: NYU Press.

United Nations. 1948. *Universal Declaration of Human Rights*. https://www.un.org/en/about-us/universal-declaration-of-human-rights.

———. 2015. 17 Sustainable Development Goals. *Transforming Our World*. https://sdgs.un.org/2030agenda.

U.N. Development Programme. 2020. *Human Development Report 2020*. New York: Oxford University Press. http://hdr.undp.org/en/2020-report.

White, Allen. 2013. Transforming the Transnational. In *Preparatory Papers Corporations in a Great Transition*. Tellus Institute.

Wright, Erik Olin. 2019. *How to Be an Anti-Capitalist in the 21st Century*. London and Brooklyn, NY: Verso Press.

Yates, Michael. 2018. *Can the Working Class Change the World?* New York: Monthly Review Press.

Index

M. T. Hill, *Building a Solidarity Society*,
https://doi.org/10.1007/978-3-031-07349-6

FSC
www.fsc.org
MIX
Papier | Fördert
gute Waldnutzung
FSC® C083411

Kolibri 360 GmbH
Ferdinand-Jühlke-Straße 7
99095 Erfurt, Deutschland
produktsicherheit@kolibri360.de